T0220521

The Second Quantum Revolution

Lars Jaeger

The Second Quantum Revolution

From Entanglement to Quantum Computing and Other Super-Technologies

 Springer

Lars Jaeger
Baar, Switzerland

ISBN 978-3-319-98823-8   ISBN 978-3-319-98824-5   (eBook)
https://doi.org/10.1007/978-3-319-98824-5

Library of Congress Control Number: 2018952603

© Springer Nature Switzerland AG 2018
This work is subject to copyright. All rights are reserved by the Publisher, whether the whole or part of the material is concerned, specifically the rights of translation, reprinting, reuse of illustrations, recitation, broadcasting, reproduction on microfilms or in any other physical way, and transmission or information storage and retrieval, electronic adaptation, computer software, or by similar or dissimilar methodology now known or hereafter developed. Exempted from this legal reservation are brief excerpts in connection with reviews or scholarly analysis or material supplied specifically for the purpose of being entered and executed on a computer system, for exclusive use by the purchaser of the work. Duplication of this publication or parts thereof is permitted only under the provisions of the Copyright Law of the Publisher's location, in its current version, and permission for use must always be obtained from Springer. Permissions for use may be obtained through RightsLink at the Copyright Clearance Center. Violations are liable to prosecution under the respective Copyright Law.
The use of general descriptive names, registered names, trademarks, service marks, etc. in this publication does not imply, even in the absence of a specific statement, that such names are exempt from the relevant protective laws and regulations and therefore free for general use.
While the advice and information in this book are believed to be true and accurate at the date of publication, neither the authors nor the editors nor the publisher can accept any legal responsibility for any errors or omissions that may be made. The publisher makes no warranty, express or implied, with respect to the material contained herein.

This Copernicus imprint is published by the registered company Springer Nature Switzerland AG
The registered company address is: Gewerbestrasse 11, 6330 Cham, Switzerland

To my friends Wolfgang and Alexander

Contents

Prologue: The White Rabbit

In Douglas Adam's parody of intergalactic life *The Hitchhiker's Guide to the Galaxy*, one reads at the beginning of the second book:

> *There is a theory which states that if ever anyone discovers exactly what the Universe is for and why it is here, it will instantly disappear and be replaced by something even more bizarre and inexplicable. There is another theory which states that this has already happened.*

The physics of the twentieth century can hardly be described more fittingly. Around 1900, physical concepts such as fields and waves, the invisible force of gravity, and entropy were already quite bizarre and difficult to grasp for a broad audience. Not all these phenomena could be seen or touched, but they were calculable and predictable and reflected what people were experiencing in their everyday lives. Despite their abstractness, they were still quite real in comparison with the mental constructs physicists had to develop to understand the nature of the atomic world (as well as the vastness of the universe).

The triumph of the totally bizarre began with the observation that at the level of atoms certain quantities cannot take just any value. For example, the radiated energy of certain bodies only assumes fixed, and in fact discrete values. It is so to speak packaged in what physicists were to call quanta (from the Latin word *quantum*—that much). If the rules of the microworld were also valid in "our" world, one would only be able set the temperature in one's living room at 10, 20, or 30 °C, while all values in between would simply not exist. A short time later, physicists realized that light has a dual nature: sometimes it is a wave, while another time it may be a particle. The same holds for the electron, as was observed shortly afterward. But how can a spatially localized particle simultaneously be a spatially extended (de-localized) wave? In the world of classical science, where white is always white and black is exactly black, this "wave-particle duality" seemed like an outrageous provocation.

By the end of the nineteenth century, physicists were just becoming accustomed to the idea that their theories would soon provide a complete understanding of everything in the world. What felt like a moment later, they were suddenly forced to say goodbye to 250-year-old physical truths and more than 2500-year-old philosophical certainties. They had to deal with more and more seemingly impossible circumstances. Quantum entities can be in several states at the same time, for example, they can be in several places at once. And then quantum entities do not even possess objectively defined properties: their properties can only be specified with probabilities, the results of measurements depend on the observer, and their quantum states (wave functions) simply decay outside any window of time. And finally there is the strangest of all quantum phenomena: the entanglement of spatially separated particles. Even when they are far apart, two particles can be coupled together as if by magic. The bottom line is that the nature and properties of quantum entities are highly abstract and can no longer be reconciled with the way we perceive and think about things in our everyday lives.

However, despite all these imponderables, today's quantum theories predict the outcome of experiments and natural phenomena with an accuracy unsurpassed by any other theory in science. Here is another counter-intuitive manifestation that contradicts any everyday experience: something that is indefinite and elusive is nevertheless a process that can be calculated 100% accurately.

> Quantum physics seems completely crackbrained. We do not understand what exactly happens, nor why, but we can calculate it precisely.

Because we can calculate ever more exactly what is going on at the atomic level, we are able to gain more and more control over the microcosm. Applications of quantum physics have long since become a concrete part of our lives. Electronics, digital technologies, lasers, mobile phones, satellites, televisions, radio, nuclear technology, modern chemistry, medical diagnostics—all these technologies are based on it. We rely on quantum technologies, even though the theory underpinning them describes a world which—in our common understanding—possesses uncertain and unsustainable features and seemingly paradoxical properties.

Only in recent years have physicists begun to realize that quantum physics can ensure a significant supply of as yet unexploited technological capabilities. The renowned quantum physicist Rainer Blatt predicts another "century of quantum technology" for the twenty-first century, enough to fundamentally change both our economy and society. We are just at the beginning of our understanding of the possibilities arising from this revolution, Blatt believes.[1]

Much of what is applied technologically has not long been fully understood theoretically, and some is still not fully understood even today. Today's quantum physicists are like magicians on a stage who pull white rabbits out of hats every evening with the utmost ease. But they have as little understanding as the audience as to how these got into the hat in the first place.

I want to take you into the completely crazy, fabulous, and incredible world of the quantum. On this journey, we will first take a look at the world of the quantum technologies that are already shaping our world today. We will then realize that we are at the beginning of another breathtaking development. In Parts II and III of the book, we take a closer look at the strange discoveries in the quantum world, which, as will be explained in Part IV, also strongly shaped the philosophical, spiritual, and religious thinking of the twentieth century. Part V then takes us to the very core of the quantum world, which at the same time represents the basis of several exciting future quantum technologies, and on which physicists have only in the last few years been able to get any kind of hold: *the phenomenon of entanglement.* Here, as we shall see, we find solutions to some of the most challenging contradictions that the founding fathers of quantum physics were struggling to resolve. However, we will encounter new questions and apparent contradictions. In the last chapter, we shall venture some suggestions as to how new quantum technologies could shape our future.

*****

Many people have read this text and made valuable suggestions for improvement. First and foremost, I would like to thank Bettina Burchardt, without whom the book would not have been possible in its original German form. For many hours, she has dedicated herself to the text and its contents and has brought this book into the form it now has. Furthermore,

---

[1] Sixty-sixth Lindau Nobel Laureate Meetings, 26 June–1 July 2016.

Michael ten Brink has made a considerable contribution to the final form of this book. I greatly appreciated his input on technical issues. Then, I would like to thank my partner Yuka Nakamura for her emotional support over the many weeks of writing and then translating. I would also like to thank Frank Wigger for his excellent project management and support during the completion of this book. Equal thanks go to my agent Beate Riess and her colleague German Neundorfer, for all their support and encouragement, not only for this book.

Despite all this help, some mistakes and many omissions seem to be inevitable. I apologize to the reader and of course take full responsibility for this.

But now, curtains up!

Baar, Switzerland
June 2018

Part I

Quantum 2.0—The Second Technological Revolution Arising from the Quantum World

1

Mighty Power: How a Theory of the Microcosm Changed Our World

It all started with three problems:

1. In 1900 Max Planck found himself unable to explain that so-called black bodies emit energy not in arbitrary quantities, but rather in "energy packets" of a certain size.
2. In 1905 Albert Einstein was forced to conclude that light is both wave and particle.
3. In 1912 Ernest Rutherford discovered in a startling experiment that the atom consists of a nucleus of protons with electrons orbiting around it; however, according to the laws of classical physics this should not be possible.

With these three phenomena in their backpacks, physicists embarked on one of the most exciting intellectual journeys in human history. In the first 30 years of the 20th century they set out from the safe shores of classical physics to cross an uncharted ocean, like the sailors of the fifteenth and sixteenth centuries, keen to find out what was on the other side.

> At the beginning of the 20th century, physicists discovered that the laws of classical physics do not always apply.

Their experiments showed them more and more clearly that some fundamental properties of the atomic world cannot be reconciled either with our everyday perceptions or with the conceptual systems of Western philosophy:

© Springer Nature Switzerland AG 2018
L. Jaeger, *The Second Quantum Revolution*, https://doi.org/10.1007/978-3-319-98824-5_1

- Superposition:
  Quantum entities can concurrently be in a mixture of different states that would be mutually exclusive in the classical world. For example, they can move simultaneously along different paths, i.e., they can be at different places at the same time.
- Randomness in behavior:
  The measurable properties of a quantum system and their temporal development can no longer be absolutely determined. With its ability to be both here and there at the same time, its observable properties can only be specified probabilistically.
- Dependence of a quantum state on measurement:
  In the micro world, measurements have a direct influence on the measured object. Even stranger is the fact that only observation assigns a definite state to a quantum particle. In essence, this means that quantum particles have no independent and objective properties. Any properties they have are obtained by an external observer.
- Entanglement:
  Quantum particles may be non-locally interconnected. Even if they are spatially far apart, they can still belong to a common physical entity (physicists say a single "wave function"). They are thus coupled together as if by some magic force.

Each of these features of the micro world violates one of four key traditional philosophical principles:

1. the *principle of uniqueness*, according to which things are in definite states (the chair stands in front of the window and not next to the door);
2. the *principle of causality*, according to which every effect must have a cause (if the chair falls over, a force must have acted on it);
3. the *principle of objectivity* (related to the *principle of reality*) according to which things have an objective existence independently of our subjective perception of them (when we leave the room, the chair remains exactly where it stands and is still there even when we no longer look at it)[1]; and

---

[1]Here, however, there had already existed important philosophical movements of thought that questioned the independence of things from our perception of them, such as the Kantian philosophy, which doubts that we can have knowledge of "things in themselves."

4. the *principle of independence*, according to which things behave individually and independently of one another (the chair is not influenced by the fact that there is another chair in the adjoining room).

For more than 2,500 years, philosophers have grappled with the existential questions of humanity. Democritus wondered whether matter could be split endlessly into smaller and smaller parts and had come to the conclusion that there must be minute particles that are indivisible, the atoms. Parmenides was in search of the ultimate and changeless substance. Aristotle and Plato were interested in how we as observers relate to the observed. There followed a hundred generations of philosophers who painstakingly sought clarity and coherent descriptions of the world. But then, at the beginning of the 20th century, it became apparent that many philosophical principles found through this tireless and thorough reflection apply only to part of the world.

> Some properties of atoms and their constituents differ completely from our everyday world of experience. Where the laws of classical physics no longer work, even philosophical principles lose their validity.

Quantum Physicists—From Magicians to Engineers

While the phenomena and properties of the micro world seemed at first like magic to physicists, with the help of mathematical means and tricks, they learned over time to calculate more and more accurately and ultimately tame this magical world, despite the fact that they did not fully understand it. Their intellectual voyage led physicists to a theory that explained the observed phenomena in the micro world, though with entirely new principles and concepts: *quantum theory*. With this theoretical basis, physicists were no longer magicians, but went back to being scientists—and later engineers, as the new theory made possible many amazing and sometimes terrifying technologies. The first of these arose when physicists applied their quantum physical models to the atomic nucleus. They realized that within in it there lay a vast amount of hidden energy.

In the years in which the world around them toppled into the chaos of two world wars and entire cities fell victim to bombing by the warring parties, physicists had to cope with the collapse of their own traditional ways of thinking. And from the bizarre new theory emerged a technology that could destroy not just a few streets, but entire cities in one fell swoop. Even as physicists—far from the public eye—were still disputing the strange and sometimes grotesque features of the micro world, quantum physics made its first appearance on the world stage, and with a very real and loud bang.

> The very first technical application of quantum physics was the most terrible weapon ever deployed by the military: the atomic bomb.

How did this terrible weapon come into existence? Since Rutherford's experiment in 1912, it had been known that the atomic nucleus consists of elementary particles carrying a positive electric charge (protons). But as we learn at school, like-charged particles repel each other. How then can atomic nuclei be stable? The many protons in the atomic nucleus should fly apart! Another force had to act attractively at the very short distances inside the atomic nucleus, and strongly enough to balance the electric force. But physicists had no idea what that force could be. Here then was another quantum puzzle!

In 1938, the German researchers Otto Hahn and Lise Meitner conducted experiments with uranium nuclei to investigate the unknown force in the atomic nucleus in more detail. The uranium nucleus contains 92 protons, and either 143 or 146 neutrons, depending on the isotope. Uranium nuclei were bombarded with slowed down neutrons and two very different elements emerged from time to time: barium and krypton. Barium atoms, which were rapidly detected by means of radiochemical techniques, have an atomic number of 56 and are thus less than half the mass of uranium nuclei. How was that possible? Using theoretical quantum physical calculations, Meitner came to the conclusion that the uranium nuclei had been broken into parts by the neutron bombardment, and the fragments absorbed a great deal of energy, much more than was the case in any previously known atomic process. But where did this energy come from? Another puzzle. Meitner also calculated that the two nuclei that emerged from the fission (along with three neutrons that were also emitted) weighed slightly less than the original uranium nucleus plus the neutron that triggered the fission. What had happened to the missing mass?

At this point, Einstein's famous formula $E = mc^2$, discovered more than 30 years earlier, entered the stage: for the difference between the total mass before and after the fission corresponded exactly to the energy that the fragments had acquired. This was the first known process in which the equivalence of energy and mass formulated by Einstein was directly revealed. At the same time, it became clear that unimaginable energies lay dormant inside these atoms!

Given the ongoing war, the presence of such a lot of energy in such a small space quickly aroused the interest of the military. In the greatest secrecy (not even the Vice President was informed), the American government put together a team of senior scientists and technicians. The goal of the *Manhattan Project*, the most complex and difficult engineering project ever undertaken until then, was the construction of an atomic bomb. The scientists were successful. On July 16, 1945, on a test site in the desert of New Mexico, the first ever atomic bomb exploded. Its force exceeded even the most optimistic expectations of the physicists. But when the immense nuclear mushroom cloud appeared on the horizon, they felt a sense of deep discomfort. As the head of the Manhattan project, Robert Oppenheimer, later reported, he quoted a phrase from the "Bhagavad Gita" of Indian mythology: "Now I am become Death, the destroyer of worlds." One of his colleagues, the director of the test, Kenneth Bainbridge, expressed it even more vividly: "Now we are all sons of bitches". Their sense of disillusionment was well justified. Only three weeks later, the second atomic mushroom emerged, this time over the skies of Japan, to be followed only two days later by the third. Just under seven years had passed from the scientific discovery of nuclear fission to the atomic mushroom clouds over Hiroshima and Nagasaki.

> With the atomic bomb, quantum physics lost its innocence right from the start. Physicists had to realize that their thirst for knowledge could destroy not only the prevailing world view, but also the world itself.

Ever More Abstract Theory, Ever More Technology—The Laser

But atomic energy also has peaceful applications, such as in nuclear power plants. And quantum physics has shaped a variety of other very useful technologies, one of which is known to all: the laser.

According to quantum theory as expounded in Bohr's atomic model, in their movements around the atomic nucleus, electrons can spontaneously jump from one orbit to another. These are the proverbial "quantum leaps". In fact, all the most important mechanisms known in nature to produce light, including chemical processes like burning, are based on such quantum leaps (radiation emitted by accelerated charged particles, such as bremsstrahlung which generates X-rays, are a relatively insignificant source of light). But how exactly do these jumps take place? In order to jump to an energetically higher state, the electron must absorb the energy of an incoming light particle (photon); when jumping back to a lower level, the electron then releases a photon. So far so good. But where exactly do light particles come from and where do they go? And yet another question arises: single quantum leaps are not causal processes that can be precisely predicted. They are instead instantaneous processes, which happen, so to speak, outside of time. What does that mean? A light switch, when activated, lights up the light, from one moment to the next. In other words, it takes a split second before the effect occurs. But when an electron jumps, no time passes, not even a fraction of a fraction of a fraction of a second.

> When an electron spontaneously jumps back to its ground level, there is no direct cause for it, and nor can we assign any definite moment or time interval at which or during which that process occurs.

In 1917, these quantum puzzles motivated Einstein to investigate the question of light absorption and emission in atoms in more detail. Planck's radiation formula describes the quantized emission of photons from black bodies. From purely theoretical considerations, Einstein succeeded in finding another, as he wrote himself, "amazingly simple derivation" of the law of spontaneous light emission. In addition, he also identified a completely new process, which he referred to as "induced light emission". This is the emission of photons from appropriately prepared ("excited") atoms, which does not occur spontaneously, but is triggered by another incoming photon. The energy thereby discharged is released into the electromagnetic field generating another photon. The triggering photon continues to exist. Thus, in an environment in which many atoms are in an excited state, i.e., many electrons are at a higher energy level, there can occur a chain reaction of electrons jumping to lower levels—and this implies a concurrent emission of light.

The interesting feature here is that each of the newly emitted photons has exactly the same characteristics: all oscillate with identical phase, propagate in the same direction, and have the same frequency and polarization (direction of oscillation). Thus, out of a few photons that initiate the chain reaction, there comes a very strong light with properties identical to those of its constituent photons. Physicists also speak of a "coherent light wave".

Only in the 1950s and 1960s did physicists succeed in and experimental proof and technological realization of this stimulated emission of photons, which Einstein had described in 1917 on purely theoretical grounds. It became the basis of the laser, another key quantum technology of the 20th century. A laser is produced in two steps: first, electrons in a medium are stimulated by light radiation, an electric current, or other processes to jump to higher energy states (physicists speak of "pumping"). Then, light particles with the same energy (frequency) as the excitation energy of the electrons are sent into the medium, causing the electrons to jump back to their ground state. They thus send out photons, which are exact copies of the incoming photons. This process gives the laser its name: *Light Amplification by Stimulated Emission of Radiation.*

Even with the laser, physicists remained in the dark for some time regarding the exact nature of the processes involved. Only an even more complex and even less comprehensible quantum theory would eventually be able to describe the atomic quantum leaps of electrons and the associated spontaneous formation and destruction of light quanta: the quantum theory of the electromagnetic field, or quantum electrodynamics. For this description, even more abstract mathematics was needed than for the original quantum mechanics.

> The laser again reveals this key feature of quantum physics: extremely abstract and non-descriptive theories can produce very real technological applications.

Quantum Physics and Electronics—From the Transistor to the Integrated Circuit

The properties of solid state of matter, such as thermal conductivity, elasticity, and chemical reactivity, are largely determined by the properties and states of the electrons in the matter. Here, too, quantum effects play a decisive role.

Among other things, quantum physics gives a precise explanation for the electrical conductivity of substances, including those of the so-called semiconductors. Their conductivity lies between those of electrical conductors (such as copper) and non-conductors (such as porcelain), but can be strongly influenced by various means. For example, changing the temperature of certain semiconductors changes their conductivity, and this in quite a different way to what happens in metals: it increases rather than decreases with rising temperature. Introducing foreign atoms into their crystal structure (a process known as "doping") can also significantly influence their conductivity. Thus, micro transistors are nothing but a combination of differently doped semiconductor elements, and their mode of operation is largely determined by the flow of electrons inside them. Once again, all this follows from the laws of quantum physics.

Semiconductor components are the building blocks of all electronics, and indeed the entire computer and information technologies that shape our lives so profoundly today. In "integrated circuits", they are packaged in the billions on small chips so that highly complex electronic circuits can be interconnected on elements as small as a few square millimeters (e.g., in microprocessors and memory chips). Today the individual elements of these integrated circuits consist of only a few dozen atomic layers (about 10 nm thick)—whatever takes place in them obeys the laws of quantum physics.

> Without making use of quantum physics, today's chips for computers, cell phones, and other electronic devices could not be produced.

An example of a quantum effect which is enormously important in microscopic transistors and diodes is the *tunnel effect*: With a certain probability, quantum particles can overcome a barrier, even if they don't strictly have enough energy for that according to the laws of classical physics. The particle simply tunnels through the energy barrier. Transferred to our macro world, that would mean that if we fired a thousand rubber arrows at a lead wall, some would appear on the other side, and what's more, we could calculate very precisely how many arrows that would be. Quantum tunneling is a bizarre feature that has very real and important consequences in today's technological world. This is due to the fact that, if the distances between the conductive regions of circuits shrink to 10 nm and less, complications arise: the electrons will tunnel uncontrollably and cause interference. To prevent this, engineers have to come up with all sorts of tricks. For example, they combine different materials so that the electrons are trapped, i.e., less likely to tunnel. In the meantime, physicists are even able to calculate the tunnel effect so well that they can construct "tunnel-effect transistors" (TFETs) whose functioning is based explicitly on the tunnel effect. Because even the "tunnel current" can be controlled.

The tunnel effect of quantum physics plays a major role in modern microelectronics—on the one hand, as an obstacle to the ever-increasing miniaturization, on the other, as the basis of a new transistor technology.

In addition to the electrical conductivity of solids, everyday properties such as color, translucency, freezing point, magnetism, viscosity, deformability, and chemical characteristics, etc., can only be understood using the laws of quantum physics. The field of solid state physics would no longer be conceivable without knowledge of quantum effects. Again and again, physicists come across surprising effects and properties and observe astonishing new macroscopic quantum effects that open the way to further possible applications.

One example is superconductivity, the complete disappearance of electrical resistance in certain metals when reduced to temperatures close to absolute zero. This effect was first observed in 1911 and can be explained by a specific many-particle quantum theory called "BCS theory", named after John Bardeen, Leon Neil Cooper, and John Robert Schrieffer who invented it in 1957. (John Bardeen thus became the only person to date to receive a second Nobel Prize in physics, in addition to the one for his discovery of the transistor effect.) However, in 1986 physicists discovered that in some materials the temperature at which they start conducting the electric current without resistance is much higher than in all previously known superconducting metals (and this was rewarded by another Nobel Prize only one year later). As is often the case in quantum physics, this phenomenon is not yet entirely understood (BCS theory does not explain it), but it has tremendous technological potential. The dream of quantum engineers is to identify substances that are superconducting at room temperature. This would allow electricity to be transported across entire countries and continents without any energy loss—about 5% of the energy in today's electricity networks actually gets lost.

New Connections—Quantum Chemistry and Quantum Biology

With quantum theory scientists also recognized a whole new connection between physics and chemistry. How atoms combine to form molecules and other compounds is determined by the quantum properties of the electron shells in those atoms. That implies that chemistry is nothing more than applied quantum physics. Only with knowledge of quantum physics can the structures of chemical bonds be understood. Some readers may recall the cloud-like structures that form around the atomic nucleus. These clouds, which are called orbitals, are nothing but approximate solutions of the fundamental equation of quantum mechanics, the Schrödinger equation. They determine the probabilities of finding the electrons at different positions (but note that these solutions only consider the interactions between the electrons and the atomic nucleus, not those between the electrons).

"Quantum chemistry" consists in calculating the electronic structures of molecules using the theoretical and mathematical methods of quantum physics and thereby analyzing properties such as their reactive behavior, the nature and strength of their chemical bonds, and resonances or hybridizations. The ever increasing power of computers makes it possible to determine chemical processes and compounds more and more precisely, and this has gained great significance not only in the chemical industry and in materials research, but also in disciplines such as drug development and agro-chemistry.

Last but not least, quantum physics helps us to better understand the biochemistry of life. A few years ago bio-scientists started talking about "quantum biology". For example, the details of photosynthesis in plants can only be understood by explicitly considering quantum effects. And among other things, the genetic code is not completely stable, as protons in DNA are vulnerable to the tunnel effect, and it is this effect that is partly responsible for the emergence of spontaneous mutations (Chap. 22 will elaborate on this further).

Yet as always, when something is labelled with the word "quantum", there is some fuzziness in the package. Theoretically, the structures of atoms and molecules and the dynamics of chemical reactions can be determined by solving the Schrödinger equation (or other quantum equations) for all atomic nuclei and electrons involved in a reaction. However, these calculations are so complicated that, using the means available today, an *exact* solution is possible only for the special case of hydrogen, i.e., for a system with a single proton and a single electron. In more complex systems, i.e., in practically all real applications in chemistry, the Schrödinger equation can only be solved using approximations. And this requires the most powerful computers available today.

> Theoretically, the equations of quantum theory can be used to calculate any process in the world.[2] However, even for simple molecules the calculations are so complex that they require the fastest computers available today, and physicists must nevertheless satisfy themselves with only approximate results.

---

[2]This statement is most likely not true on a cosmic scale. Here the general theory of relativity applies, and this theory has so far proved to be incompatible with any quantum theory (see Chap. 14).

Quantum Physics Everywhere—And Much More to Come

From modern chemistry to solid state physics, from signal processing to medical imaging systems—today we encounter quantum physics every-where. We trust its laws when we get into a car (and rely on on-board elec-tronics), power up our computer (which consists of integrated circuits, i.e., electronics based on quantum phenomena), listen to music (CDs are read by lasers, a pure quantum phenomenon), undergo X-ray or MRI scans of our bodies,[3] let ourselves be guided by GPS, or communicate via mobile phone. According to various estimates, between one-quarter and one-half of the gross national product of industrialized nations today is directly or indirectly based on inventions that have their foundation in quantum theory.

This percentage will increase rapidly in coming years. In the wake of nuclear technology, medical applications, lasers, semiconductor technology, and modern physical chemistry, all developed between 1940 and 1990, a second generation of quantum technologies has started to emerge over the past 25 years, and this is likely to shape our lives even more dramatically than the first generation. This has also been recognized by a country that was long regarded as a developing country when it came to scientific research, but has meanwhile been catching up with huge strides: the People's Republic of China. In its 13th Five-Year Plan, it has specified the new quantum tech-nologies as a strategic area of scientific research. Meanwhile, Europe has also recognized the signs of the times and has begun investing massively in quan-tum technologies. These will be the subject of the next three chapters.

> More than 100 years ago the first quantum revolution began to take shape. We are now experiencing the beginning of the second quantum revolution.

---

[3]There are two types of X-ray radiation: *Bremsstrahlung* and characteristic radiation. For the explanation and application of the former, classical physics is sufficient. It was discovered by Konrad Röntgen, who received the very first Nobel Prize in Physics in 1901. The second requires quantum physics and was discovered by Charles Glover Barkla, who received the 1917 Nobel Prize in Physics.

2

There's Plenty of Room at the Bottom: A New Generation of Quantum Technologies

In 1959, the quantum physicist and later Nobel laureate Richard Feynman gave a much-cited lecture which outlined how future technologies could operate on a micro- and nanoscopic scale (scales of one thousandth or one millionth of a millimetre, respectively). The talk was entitled "*There's Plenty of Room at the Bottom*". Feynman's vision was very concrete: he predicted that man would soon be able to manipulate matter down to the level of individual atoms. Feynman's lecture is considered to be the big bang of nanotechnology, one of the most exciting technologies being developed today. Its goal is the control and manipulation of individual quantum states.

In fact, many of Feynman's ideas have already long since become a reality. Examples are:

- The electron microscope, in which the object to be observed is scanned point by point using an electron beam with wavelength up to 10,000 times shorter than the wavelength of visible light. This allows resolutions up to 50 pm (50×10^{-12} m) and magnifications up to 10,000,000, while light microscopes cannot achieve resolutions of more than 200 nm (200×10^{-9} m) and magnifications of 2,000.
- Microscopic data storage units based on semiconductor technology, which allow 500 gigabytes to be stored on a thumbnail-sized surface.
- Integrated circuits with elements involving only 10 to 100 atoms each, which enable the ultrafast processing of information in modern computers, thanks to the vast numbers of them that can be integrated into a single microchip.

© Springer Nature Switzerland AG 2018
L. Jaeger, *The Second Quantum Revolution*, https://doi.org/10.1007/978-3-319-98824-5_2

- Nanomachines in medicine, which can be introduced into the human body, e.g., to search autonomously for cancer cells.

Many of Feynman's visions from 1959 are already part of our everyday technological lives today.

Feynman's most ground-breaking vision in 1959, however, concerned the possibility of constructing ultra-small machines that can manipulate matter at the atomic level. These machines would be able to put together any kind of material from a kit of atoms of various elements, rather like playing Lego using a manual given by humans, the only prerequisite being that the synthetically produced composites must be energetically stable.

First versions of such basic building blocks exist already: nano wheels that can actually roll long, nano gearwheels that spin along a jagged edge of atoms, nano propellers, hinges, grapples, switches, and more. They are all about ten thousandths of a millimetre in size, and obey the laws of quantum physics rather than classical Newtonian mechanics, which makes nanotechnology essentially a quantum technology.

In his science fiction novel "The Lord of All Things" (in German. "Herr der kleinen Dinge", 2011), the science fiction author Andreas Eschbach describes how nanomachines can put together individual atoms and molecules in almost any desired way. Eventually, they start replicating themselves in a way which leads to an exponential expansion in their numbers. Thanks to their abilities, these nanomachines are able to produce things almost out of nothing. The main protagonist of the novel learns to control them and has them spontaneously build things he needs at any particular moment (cars, planes, even a spaceship). Ultimately, he manages to control these processes solely through his own thoughts, by having them directly measure his brain signals.

Are such nanomachines actually possible or is this pure science fiction? Feynman claims that there is no law of nature that speaks *against* their construction. In fact, today's nano-researchers are getting closer and closer to his vision. The 2016 Nobel Prize for Chemistry, awarded to Jean-Pierre Sauvage, Fraser Stoddart, and Bernard Feringa for their work on molecular nanomachines, shows how important the research community considers this particular work.

The nanomachines predicted by Richard Feynman, which can seemingly assemble (almost) any material out of nothing from raw atomic material, or repair existing—even living—material, are theoretically possible. The first steps toward such machines have already been taken and they are likely to do much to shape the 21st century.

Quantum Spookiness Becomes Technology

In a second visionary speech in 1981, Feynman developed what is perhaps an even more radical idea: a whole new kind of computer, called a "quantum computer", which would make today's high-powered computers look like the Commodore 64 from the early 1980s. The two main differences between a quantum computer and today's computers are:

- In the quantum computer, information processing and storage no longer occur by means of electron currents, but are based on the control and steering of single quantum particles.[1]
- Thanks to the quantum effect of superposition, a quantum computer can calculate on *numerous* quantum states, called quantum bits (qubits), at the same time. Instead of being constrained to the states 0 and 1 and processing each bit separately, the possible states that can be processed in one step are thereby multiplied in a quantum computer. This allows an unimaginably higher computing speed than today's computers.

While quantum computer technology is still in its infancy, when it reaches adulthood it will dramatically speed up a variety of algorithms in common use today, such as searching databases, computing complex chemical compounds, or cracking common encryption techniques. What's more, there are a number of applications for which today's computers are still not powerful enough, such as certain complex optimizations and even more so potent machine learning. A quantum computer will prove very useful here. And at this point the quantum computer will meet another ground-breaking future technology: the development of artificial intelligence. Quantum computers will be the subject of Chap. 4.

---

[1]There exist a variety of different concepts for building quantum computer (see Chap. 4). Some of these actually use entire ensembles of particles, but ensembles which behave like single quantum particles.

In quantum physics, Richard Feynman no longer saw just the epitome of the abstract, but very concrete future technological possibilities—this is what Quantum Physics 2.0 is about.

As Feynman predicted almost 60 years ago, we already use a variety of quantum-physics-based technologies today. Common electronic components, integrated circuits on semiconductor chips, lasers, electron microscopes, LED lights, special solid state properties such as superconductivity, special chemical compounds, and even magnetic resonance tomography are essentially based on the properties of *large ensembles of quantum particles* and the possibilities for controlling them: steered flow of *many* electrons, targeted excitation of *many* photons, and measurement of the nuclear spin of *many* atoms. Concrete examples are the tunnel effect in modern transistors, the coherence of photons in lasers, the spin properties of the atoms in magnetic resonance tomography, Bose–Einstein condensation, or the discrete quantum leaps in an atomic clock. Physicists and engineers have long since become accustomed to bizarre quantum effects such as quantum tunnelling, the fact that many billions of particles can be synchronized as if by magic, and the wave character of matter. For the statistical behaviour of an ensemble of many quantum particles can be well captured using the established quantum theory given by Schrödinger's equation, now 90 years old, and the underlying processes are still somewhat descriptive. They constitute the basis of the first generation of quantum technologies.

The emerging second generation of quantum technologies, on the other hand, is based on something completely new: the directed preparation, control, manipulation, and subsequent selection of states of *individual quantum particles* and their interactions with each other. Of crucial importance here is one of the strangest phenomena in the quantum world, which already troubled the founding fathers of quantum theory. This is *entanglement*, which will be the subject of the entire fifth part of the book (Chaps. 21, 22, 23, 24, 25 and 26).

Previous quantum technologies were essentially based on the behaviour of many-particle quantum systems. The next generation of quantum technologies has its foundation in the *manipulation of the states of single quantum particles*.

With entanglement, precisely that quality of the quantum world comes into focus which so profoundly confused early quantum theorists such Einstein, Bohr, and others, and whose fundamental significance physicists did not fully recognize until many years after the first formulation of quantum theory. It describes how a finite number of quantum particles can be in a state in which they behave as if linked to each other by some kind of invisible connection, even when they are physically far apart. It took nearly fifty years for physicists to get a proper understanding of this strange phenomenon of the quantum world and its violation of the locality principle, so familiar to us, which says that, causally, physical effects only affect their immediate neighbourhoods. To many physicists it still looks like magic even today. No less magical are the technologies that will become possible by exploiting this quantum phenomenon.

In recent years, many research centres for quantum technology have sprung up around the world, and many government funded projects with billions in grants have been launched. Moreover, high tech companies have long since been aware of the new possibilities raised by quantum technologies. Companies like IBM, Google, and Microsoft are recognizing the huge potential revenues and are thus investing heavily in research on how to exploit entangled quantum states and superposition in technological applications. Examples include Google's partnerships with many academic research groups, the Canadian company D-Wave Systems Quantum Computing, and the investments of many UK companies in the UK National Quantum Technologies Program.

In May 2016, 3,400 scientists signed the *Quantum Manifesto*, an initiative to promote co-ordination between academia and industry to research and develop new quantum technologies in Europe.[2] Its goal is the research and successful commercial exploitation of new quantum effects. This manifesto aimed to draw the attention of politicians to the fact that Europe is in danger of falling behind in the research and development of quantum technologies. China, for example, now dominates the field of quantum communication, and US firms lead in the development of quantum computers. This plea has proved successful because the EU Commission has decided to promote a flagship project for research into quantum technologies with a billion euros over the next ten years. That's a lot of money given the chronically weak financial situation in European countries. The project focuses on

---

[2]*Quantum Manifesto—A New Era of Technology*, available from http://qurope.eu/system/files/u7/93056_Quantum%20Manifesto_WEB.pdf.

four areas: communication, computing, sensors, and simulations. The ultimate goal is the development of a quantum computer.

> The EU is funding a dedicated project on quantum technologies with one billion euros over ten years. Politicians have high expectations from this area of research.

No wonder such a lot of money is being put into this field of research, as unimaginable advantages will reward the first to apply and patent quantum effects as the basis for new technologies. Here are some examples of such applications, the basics of which physicists do not yet fully understand (apart from superconductivity, already mentioned in the last chapter):

- The quantum Hall effect discovered in the 1980s and 1990s (including the fractional quantum Hall effect). These discoveries were rewarded by Nobel Prizes in 1985 and 1998, respectively. This states that it is not only energy that is emitted in packets, but at sufficiently low temperatures, the voltage that is generated in a conductor carrying an electric current in a magnetic field (classic Hall effect) is also quantized. This effect makes possible high-precision measurements of electric current and resistance.
- New miraculous substances such as graphene, which are very good conductors of electricity and heat and are at the same time up to two hundred times stronger than the strongest type of steel (Nobel Prize 2010). Graphene can be used in electronic systems and could make computers more powerful by several orders of magnitude.
- Measuring devices based on the fact that even very small forces, such as they occur in ultra-weak electric, magnetic, and gravitational fields, have a quantifiable influence on the quantum states of entangled particles.
- Quantum cryptography, which is based on the phenomenon of particle entanglement (Nobel Prize 2012) and allows absolutely secure encryption.

By considering the last two examples, we shall show what dramatic effects the new quantum technologies, even apart from the quantum computer, may have on our everyday lives.

Measuring Ever More Accurately—Possible through new Quantum Technologies

The accurate measurement of physical quantities like how far New York is from Boston or how many electrons flow through a given wire in a given time may sound pretty boring. But it is not. Because no matter what is being measured, be it meters, seconds, volts, or whatever, the highest accuracy may be crucial. The sensitivity of quantum mechanically entangled states to external disturbances can be extremely useful in this respect for various measurement purposes.

A well-known example of the metrological application of quantum physical processes is the measurement of time by atomic clocks. Optical atomic clocks have been around for 70 years now. They receive their time interval from the characteristic frequency of electron transitions in atoms that are exposed to electromagnetic radiation. The commonly used caesium atoms have a maximum resonance for incoming electromagnetic waves with a frequency of 9,192,631,770 oscillations per second (in the microwave range), i.e., at that frequency, a maximum of photons is emitted. With the widely accepted definition that one second equals 9,192,631,770 of these vibrations, humans have a much more accurate definition of the second than the statement that one day contains 86,400 s.

Atomic clocks are particularly accurate because they are based on the stimulation of many caesium atoms and a mean value of the number of emitted photons is taken. The measurement becomes even more accurate now that there are around 260 standardized atomic clocks worldwide that can be compared with each other, giving rise to yet another averaging effect.

> Time measurement is unimaginably accurate, thanks to a worldwide network of atomic clocks. They are accurate to within 1 s every million years.

And yet, that is not accurate enough. How can that be? After all, our clock only has to be accurate to the nearest second to be sure that we don't miss the beginning of our favourite TV show. But what most of us don't take into account is that the global navigation system GPS would not work without atomic clocks, because they determine positions by means of a measurement of the time a signal takes to be transmitted between the device and the GPS satellites. In order to be able to determine our position to within a meter, the time measurement must be accurate to a few billionths of a second.

Likewise, digital communication in which large numbers of telephone calls are transmitted simultaneously over a *single* line depends on ultraprecise time measurement. Atomic clocks control the switches that route the individual digital signals through the network in such a way that they arrive in the right order at the right receiver.

The accuracy of atomic clocks can be affected by external disturbances such as electric fields. These broaden the frequency spectrum of the measured photons, leading to small deviations in the resonance frequency and thus also in the time measured. Another influence comes from fluctuations in the terrestrial magnetic field. This limits the accuracy of today's GPS and digital communications technology, not to mention high-precision measurements in physics experiments.

> Even with atomic clocks, the measurement of time remains too inaccurate for some applications in GPS or the multiple use of data communication channels.

A new generation of atomic clocks that take advantage of the effect of quantum entanglement would remedy this shortcoming. A few of the atoms in each clock within the global network would be quantum mechanically entangled. In this way, the clocks would stabilize each other, because a measurement on a single atom of one clock would at the same time be a measurement on all others; due to the nature of entanglement, even the smallest deviations within the network of clocks would be immediately corrected.

There is yet another way to increase the accuracy of atomic clocks through quantum physical processes. If the disturbing magnetic field fluctuations were known for every fraction of a second, we could account for them by applying an appropriate error correction scheme. Nature itself shows us how the magnetic field can be measured ultra-precisely using the effect of quantum entanglement at the atomic level (see also Chap. 22).

Many migratory bird species have a magnetic sense that they use for orientation during their flights—many of them traveling thousands of miles to their wintering grounds. The precision with which they measure the strength and direction of the Earth's magnetic field amazed ornithologists for a long time. Only a few years ago, they realized that birds use a kind of quantum compass for this purpose. In the eye of the robin, electron pairs are entangled across two molecules by their spins. These entanglements are very sensitive to external magnetic fields. Depending on the orientation of the magnetic field, the electrons rotate in different directions which corresponds to different orientations of their "spin" (more about spin in Chap. 10). In these particular molecules in the bird's eye, the change in the orientation of the electron spins is enough to transform them into isomers (molecules with the same chemical formula but different spatial structure). The different properties of the isomers depend very sensitively on the strength and direction of the magnetic field, causing different chemical reactions that eventually lead to perception in the bird's retina—the bird's eye thus becomes an ideal measuring device for magnetic fields.

> Evolution has equipped many species of birds with a kind of quantum pair of glasses for magnetic fields. With the help of quantum effects, they can thus find their way to their winter quarters.

Apart from time and magnetic fields, local gravitational fields can also be measured very accurately using quantum mechanically entangled states, something that has stimulated significant commercial interest. Today, metal and oil deposits in the ground are being located by precise measurements of the strength of local gravitational fields. Local density variations, and therefore a correspondingly slightly stronger or weaker gravitational force, also provide evidence for large underground gas or water fields—but this is a tiny effect that can only be detected with ultra-sensitive gravity sensors. By exploiting the phenomenon of quantum mechanical entanglement, such measurements could be made even more accurate. With an entanglement-based ultra-sensitive gravity sensor, even a single person could be tracked down just by the gravitational field created by their body mass. Gas pipes in the ground, leaks in water pipes, sinkholes under roads, or irregularities under the plot for a planned house could all be detected. Furthermore, the job of archaeologists could be dramatically simplified if they were in a position to simply "light up" historic and prehistoric sites with the help of such gravity sensors.

In addition, entanglement-based measurement devices could gauge the tiny magnetic currents associated with our brain activity or the cell-to-cell communication in our bodies. They would make it possible to monitor individual neurons and their behaviour in real time. This would allow the processes in our brain (and our body) to be measured much more accurately than through today's EEG recordings. Today, quantum magnetic field sensors are already used for magnetoencephalography (MEG), with which the magnetic activity of the brain can be measured by means of Superconducting Quantum Interference Devices or SQUIDs (superconducting quantum interference units). Maybe someday we could even capture our thoughts from the outside and feed them directly into a computer. Indeed, future quantum technologies could provide the perfect brain–computer interface.

> Measurement devices based on entanglement of quantum particles will make visible much of what has so far remained invisible.

The Holy Grail of Data Security—Quantum Cryptology

Let us take a deeper look at the second example from the list above: quantum cryptology. Data security is a topic that has become increasingly important in the modern world. How can we ensure that strangers do not have access to our private digital data? Or that third parties do not overhear our conversations without permission? Conventional encryption relies on encoding the message with a key code in such a way that decrypting without knowledge of the key would require unattainably high computational capacities. But this is like a never-ending race: more and more complicated encryption codes must be developed to ensure that increasingly powerful computers cannot crack them. Quantum cryptography provides a way around this, at least for the problem of the unrecognized eavesdropper.

An essential component of quantum-secure communication is *quantum key distribution*: this method of transmitting the key with entangled quantum states of light makes any intervention in the transmission, such as an eavesdropper in the communication channel, immediately detectable to the user.

Suppose A calls B on a "secure" cell phone (in quantum cryptography, A and B are always taken to stand for Alice and Bob). Alice's and Bob's devices can each take measurements on entangled particles. When the line is intercepted, Alice and Bob immediately notice that an unwanted third party (often called Eve) is on the line, because Eve would irretrievably destroy the entanglement of the particles while listening in, i.e., measuring it for that purpose. Nor can she just copy them and send the information, the qubit, to the actual addressee without being detected, because it is impossible to replicate any (not yet measured) quantum state identically (this is the *No Cloning Theorem*, see Chap. 25 for more details). As soon as Alice and Bob notice any changes to their key, or indeed that the entanglement of their particles has been destroyed, they quickly change the mode of communication and thereby thwart the eavesdropper, at least for a certain time.

Cryptology makes use of a fundamental principle of the quantum world: quantum states can never be duplicated without the corresponding state or original information changing.

While the bizarre characteristics of the micro world caused so much confusion among physicists in the first half of the 20th century, engineers are now working to apply these same features, having taken a break for a few decades. During the development of the first generation of quantum technologies, physicists once again went back to the theoretical drawing board to get a proper understanding of the laws that govern the micro world. And in the meantime, they have made significant progress in these efforts. The path is now is open to apply quantum physics and *all its key features* in a technological context. The interesting thing about this process is that scientists and engineers are not only trying to make existing and familiar things faster or more precise—instead, they are working on a whole new world of possibilities never imagined before.

As early as 1997, the physicist Paul Davies wrote: "The nineteenth century was known as the machine age, the twentieth century will go down in history as the information age. I believe the twenty-first century will be the quantum age".[3]

---

[3]Preface to G. J. Milburn, *Schrodinger's Machines: The Quantum Technology Reshaping Everyday Life*, W. H. Freeman (1997).

3

Technology on the Smallest Scales: The Possibilities of Nanotechnology

We already use it in many different ways, but only very few people are aware of it: nanotechnology. In addition to the quantum computer (the topic of the next chapter), nanotechnology offers the most exciting future technological applications of quantum theory. Many of its applications are already integrated into our everyday lives. Some examples are:

- Sun cream lotions, in which nanotechnology provides protection against UV rays.
- Nanotechnologically treated surfaces used for self-cleaning window panes, scratch-resistant car paint, and ketchup that flows steadily out of the bottle.
- Nano-treated textiles that prevent the smell of sweat. Antibacterial silver particles, for example, prevent bacteria from decomposing our odourless sweat into unpleasant-smelling bodily whiff.

Even more astounding are the emerging nanotechnologies. In Chap. 2, there was already talk of nano-robots that automatically and permanently detect pathogens in our bodies, and autonomous nanomachines that can produce just about anything from a pile of soil.

> Nanotechnology has long become indispensable in our daily lives—but the future belongs even more to this technological offshoot of quantum physics.

© Springer Nature Switzerland AG 2018
L. Jaeger, *The Second Quantum Revolution*, https://doi.org/10.1007/978-3-319-98824-5_3

One could get the impression that everything exciting and futuristic has to do with "nano". But what exactly is nanotechnology?

Nano—infinite Possibilities on the Scale of the Invisibly Small

The term "nanotechnology" was first -defined in 1974 by Norio Taniguchi:

> Nano-technology mainly consists of the processing of, separation, consolidation, and deformation of materials by one atom or one molecule.[1]

The term "nano" refers to the properties of particles and materials in the range of one nanometre to 100 nm (1 nm is one millionth of a millimetre). For comparison, the DNA double helix has a diameter of 1.8 nm, a soot particle, about 2,000 times smaller than the full stop at the end of this sentence, is 100 nm in size. The structures of the nanocosm are thus significantly smaller than the wavelengths of visible light (about 380–780 nm).

Three features make the range of the nano very special:

- It is the frontier between the world of atoms and molecules, where quantum physics applies, and the macro scale with its classical laws. In this intermediate area, scientists and engineers can make particular use of quantum effects to prepare materials with unique properties. This includes, for example, the tunnel effect, which (as mentioned in the first chapter) plays an important role in modern transistors.
- When combined with other substances, nanoparticles collect a large number of other particles around them, and this is very useful, for example, for scratch-resistant automotive paints. If a crack appears in the material, the nanoparticles fix the paint by behaving like elastic rubber bands.
- Generally speaking, surface atoms are more easily torn away from atomic compounds, which makes nanoparticles good catalysts for chemical reactions. A simple geometric consideration illustrates this. A cube with side one nanometre (about 4 atoms) contains on average of 64 atoms, 56 of which (87.5%) are located on the surface. The larger the particle, the fewer surface atoms available for reactions relative to bulk atoms. In a

---

[1]N. Taniguchi, *On the basic concept of nanotechnology.* In: *Proc. Intl. Conf. Prod. Eng. Tokyo, Part II, Japan Society of Precision Engineering* (1974).

nanocube with side 20 nm (containing 512,000 atoms), only 7.3% of the atoms are located on the surface. For side 100 nm, their number drops to 1.2%.

> Nanoparticles consist almost entirely of surface, which makes them highly reactive and gives them unexpected mechanical, electrical, optical, and magnetic properties.

In (quantum) theory, this has long been clear to physicists. However, there have not always been the tools needed to actually isolate and process matter on the nano-scale.

A ground-breaking event for nanotechnology was the development of the Scanning Tunnelling Microscope (STM) by Gert Binning and Heinrich Rohrer in 1981 (for which they were awarded the 1986 Nobel Prize in Physics). This device allows the observation of single atoms. Due to a special quantum effect (the tunnelling effect), the electric current between the tip of the grid and the electrically conductive sample reacts very sensitively to changes in their separation as low as one hundredth of a nanometre. In 1990 Donald Eigler and Erhard Schweizer even succeeded in transporting individual atoms from A to B by making specific changes to the voltage applied to the STM grid tip—the device thus no longer only just *observed* individual atoms, but also *manipulated* them. The two researchers "wrote" the logo of their employer IBM with 35 xenon atoms on a nickel crystal. Twenty-two years later, researchers were able to build a one-bit memory cell from just 12 atoms (today's standard one-bit memory cells still contain hundreds of thousands of atoms).

> What Feynman proposed in 1959 as a vision of the future, namely the construction of extremely tiny products atom by atom, is within reach today.

Physicists and engineers are not only aiming to manipulate atoms and design tiny components, they are also developing new materials (and understanding old ones) thanks to quantum physics.

Wonder Materials

For 2,000 years, skilled blacksmiths have produced the coveted Damascus steel in an elaborate manufacturing process. Layers of various steels are stacked, forged together, repeatedly folded over and flattened again and again, rather as the baker kneads dough until a material consisting of up to several hundred of these layers is finally produced. Compared to normal steel, Damascus steel is extremely hard and at the same time extremely flexible. It is known today that these extraordinary material properties are, among other things, due to the inclusion of carbon nanotubes of up to 50 nm in length and 10 to 20 nm in diameter. Of course, the ancient and medieval blacksmiths knew nothing of nanotubes, as their techniques were based entirely on experience.

As further examples, humans were already producing sparkling surfaces of metallic nanoparticles on ceramics in Mesopotamia and Egypt 3,400 years ago, while the Romans used nanoparticles to seal their everyday ceramics, and in the Middle Ages, glass containing gold nanoparticles was the recipe used to make red stained glass windows.

> Materials with properties based on nanoparticles have been manufactured and used since ancient times.

With the understanding offered by quantum physics, we can today understand and even improve materials such as Damascus steel. Through carefully specified addition of certain elements, millennia-old forging processes can be further perfected. For this purpose, nanometre-sized nickel, titanium, molybdenum, or manganese particles can be incorporated with perfect precision into the iron crystal lattice of steel. In particular, nickel and manganese promote the formation of nanocrystals which retain their structure when the metal is deformed, thus providing the material's stability. At the same time, the steel becomes very flexible and deformable due to the fine distribution of these nanocrystals. Although they make up only a very small fraction of the total mass, the additional particles bestow highly improved properties over the pure iron crystal lattice. This approach is used, for example, in the automotive and aircraft industry, in which ever more deformable and at the same time more resilient steels allow in particular lightweight material- and energy-saving construction methods.

Numerous methods for the production of nanomaterials are based on the principle of working super-fine distributions of nanoparticles into materials (which is called "doping" in semiconductors).

The "seasoning" of materials with single atoms or nano-atomic compounds can provide the original material with completely new properties, allowing us to make:

- foils that conduct electricity,
- semiconductors with precisely controlled characteristics (the basis of computer technology for decades now),
- creams that filter out the UV components from sunlight.

Another application of nanotechnology is to recreate products that nature has come up with. Spider silk is a thin thread only a few thousandths of a millimetre thick, but extremely ductile, heat-resistant up to 200 degrees, and five times as tear-resistant as steel. For decades, researchers have dreamed of producing such a substance in the laboratory. Now this dream has come true. The secret of natural spider's thread lies in a combination of chain-shaped proteins and short pieces of carbohydrate, with lengths in the nanometer range.

Artificial spider silk can be used to create super-textiles contributing to explosion-resistant gear for soldiers, super-elastic clothing for athletes, and encasements for breast implants that avoid painful scarring.

Evolution produced and used nanomaterials long before we did. Thanks to the findings of quantum physics, we can rebuild and even refine these today.

More Valuable Than Diamonds

There also exist pure nanomaterials. An interesting example is graphite. Graphite is a form of elementary carbon, used to make pencil leads, for example. It is nothing more than a stack of carbon layers, each as thick as a single carbon atom. Each layer is a two-dimensional carbon molecular lattice called *graphene*, governed by the laws of quantum physics.

Scientists have been studying these ultra-thin carbon layers theoretically for many years. Their quantum-physical calculations and models suggested that graphene must have some amazing properties: 200 times stronger than steel, an excellent electrical and thermal conductor, and transparent to visible light. They only lacked the practical proof that their theoretical calculations were correct.

Then, in 2004, Andre Geim and Konstantin Novoselov succeeded in isolating pure graphene. Their trick was to remove it with a kind of sticky tape from graphite. Geim and Novoselov received the Nobel Prize in Physics for this in 2010. Has there ever been a Nobel Prize in Physics awarded for something that simple?

With thicknesses of the order of one nanometre, graphene is the thinnest material in the world. At the same time, its atoms hold together firmly, because they are all interconnected by tightly arranged "covalent" chemical bonds. There are, so to speak, no weaknesses in this material, no places where it could break. Because in this composite each carbon atom is available for chemical reactions on two sides, it has extraordinary chemical, electronic, magnetic, optical, and even biological properties. Some possible applications of graphene are:

- Production of clean drinking water: graphene membranes can be used to build enormously efficient desalination plants.
- Energy storage: graphene can be used to store electrical energy more efficiently and more durably than other materials; long-lasting and at the same time lightweight batteries can be made.
- Medicine: scientists are doing research on artificial retinas made of graphene (see below).
- Electronics: the world's smallest transistor is made of graphene.
- Special materials: graphene could also be used as a coating which allows for the construction of flexible touchscreens—mobile phones could then be worn like bracelets.

The EU considers the prospects for graphene based technologies so solid that in 2013 it declared research in this field as one of two projects in the *Future and Emerging Technologies Flagship Initiative*, each funded with one billion euros. (The other funded project is the Human Brain Project, but meanwhile a third has come along: the flagship project on quantum technologies mentioned in the last chapter.)

The nanomaterial graphene is considered to be a future wonder material.

From Micro- to Nanoelectronics

Modern microelectronics is based on a nanomaterial: doped silicon crystals. We have thus long been following the road from micro- to nanoelectronics. And some parts of Feynman's vision have already been achieved. In 1959, he said that the contents of 25 million books could be stored in a speck of dust. To do this, one bit must be stored in 100 atoms. Today, elementary storage units with 12 atoms are possible. So on a speck of dust there is room for almost 250 million books.

An example of future nanomaterials in electronics are carbon nanotubes, also just called nanotubes. These are graphene layers rolled into tubes to make tiny carbon cylinders with a diameter of about 100 nm. Their special electrical properties can only be explained by the laws of quantum physics. Depending on the diameter of the tube, they conduct electronic currents better than any copper conductor, because the electrons move through the tube virtually without interference, i.e., without being deflected by obstructing atoms as they would be in a metallic conductor.

Researchers at Stanford University have constructed a working computer with 178 nanotube transistors.[2] It has the computing power of a computer from 1955 which would have filled an entire gymnasium.

The nanomaterial "silicene" goes even further. Here, as for graphene, atoms are arranged in two-dimensional layers with a honeycomb patterns. But while graphene is made of carbon, silicene is a foil made from elementary silicon, a semiconductor, and this makes it especially interesting for the construction of computer chips. In 2014 researchers at the University of Texas built the first transistor made of silicene. Although the production and processing of silicene still involves great technical difficulties (for example, it decays when exposed to oxygen), there is great hope that this material can significantly increase the performance of computer chips.

> Transistors based on nanotubes or silicon could be switched much faster, making the corresponding computer chips much more powerful.

However, the development of nanotubes for use in computers is not yet the end of the story. The ultimate goal of physicists and computer build-

---

[2]M. Shulaker et al., *Carbon nanotube computer*, Nature 501, 526 (26 September 2013).

ers is to use *single molecules* as transistors. In fact, certain organic molecules can already be converted from electrically conductive to insulating, just by throwing a switch.

Ultra-Small Machines—Masters of the Nano-World

Our ever more reliable technological mastery of the nano-world will foster many more technological possibilities, including Feynman's vision of ultra-small machines doing their work right at the level of single atoms. Nanowheels, nanomotors, and even a nano-elevator have already been developed, and there is a nano-car with four separate motors mounted on a central support, where the tip of a scanning tunnelling microscope fuels the moving molecule with electricity to set it in motion.

Nano-engineers can make things even tinier. The smallest electric motor in the world is just a nanometre in size and consists of a *single* bent thioether molecule sitting on a copper surface. In this molecule, two hydrocarbon chains of different lengths (a butyl and a methyl group) hang like little arms on a central sulphur atom. The whole molecule is linked to the copper surface in such a way that it can rotate freely. It is driven by a scanning tunnelling microscope whose electrons excite the molecule's rotational degrees of freedom with the help of the tunnel effect. The running speed of the motor can be controlled by the electronic current and the outside temperature.[3]

> Nanomachines are already being built today. The molecular motor today is at the same level of development as the electric motor in the 1830s.

Nobody suspected in 1830 that the electric motor would ever drive trains, dishwashers, and vacuum cleaners. The Nobel Prize Committee in Stockholm anticipated a similar potential for molecular nanomachines when deciding on its 2016 prize for chemistry. It is likely that molecular motors will soon be used in sensors, energy storage systems, and the development of new materials.

---

[3]C. Sykes et al., *Experimental demonstration of a single-molecule electric motor*, Nature Nanotechnology 6, S. 625 (2011).

Largely outside public attention, nanotechnology has taken some significant steps:

- The first generation of nanotechnology products were still passive materials with certain well-defined properties that did not themselves change when used, e.g., Damascus steel.
- In the second generation, nanotechnology produced tiny machines that "do work"—in other words, they drive an active process, e.g., a transport vehicle for targeted drug delivery in the body (see below). Now nanostructures interact and react directly with other substances, thereby changing themselves and/or their environment.
- And a third generation of nanotechnologies is already emerging: "integrated nano-systems". Here, various active nano-components such as copiers, sensors, motors, transistors, etc., are used as components and assembled into a functioning whole, much like an engine, a clutch, electronics, tires, etc., becoming an automobile when integrated. This paves the way for increasingly complex nanomachines.

> The next step in nanotechnology is to couple nanostructures with different properties and capabilities into complex nanomachines.

When Nanotechnology and Biotechnology Merge

Sixty years ago, Richard Feynman realized that nanoparticles and nanomachines could also be of great importance in medicine. This part of his vision is also becoming a reality today. Here are three examples that are already being implemented:

- The Israeli company "Nano-Retina" has developed an artificial nano-retina that allows the blind to see again[4]: it consists of a tiny, flat implant containing a high-resolution network of nano-electrodes. The nano-retina stimulates the optic nerve in such a way that incoming light particles collected by the electrodes are transmitted as visual stimuli to the brain.

---

[4]S. Roux et al., *Probing the functional impact of sub-retinal* prosthesis, eLife 2016;5:e12687 https://doi.org/10.7554/elife.12687 (23 August 2016), https://elifesciences.org/articles/12687.

- Lab on a chip: nano biosensors detect antibodies and specific enzymes in human body fluids. Only one-thousandth of a millilitre of blood, urine, or saliva (or even less) is placed on a credit card-sized chip. The nanoparticles integrated on it detect characteristic chemical, optical, or mechanical changes upon contact with the targeted ingredient. Thus, the chip can provide diagnoses for numerous disease symptoms in just a few minutes.
- Nanoparticles transport drugs directly to sites of inflammation or mutant cells to provide an efficient attack with those with drugs. For a long time, the question of how to transport nanostructures in the blood remained unresolved, because blood proves to be as viscous as honey for such small particles. Now they can even be steered, for example, by magnetic fields. Among other things, bio-engineers plan to use them in precise chemo-therapies against cancer cells.

Ultra-small nano-robots, also called "nanobots", raise enormous hopes in medicine. Our health checkup at the doctor's every two years would be replaced by a perpetual nano-check. Nanobots would wander permanently through our bodies and preventively detect pathogens, gene mutations, and dangerous deposits in the bloodstream. They would then begin the therapy immediately by delivering drugs directly to the site of the disease. They would fight viruses, inhibit inflammation, remove cysts and cellular adhesions, and prevent strokes by opening blocked arteries, and even perform surgical procedures. If necessary, they would send the results directly to the family doctor, who would then call the patient for an appointment.

> Medics have a vision of many tiny nano-robots—biomarkers, labs-on-a-chip, and other telemedical devices—permanently circulating inside our bodies for the purposes of health care and healing.

Nanoparticles or nanobots could also be used in our diet. They would help us digest food in such a way that nutrients are optimally absorbed by our body. This would be helpful for treating diseases that today require a strict diet. Researchers are also working to produce foods with nanoparticles on the surface which are designed to give our taste buds a great taste of chips, chocolates, or gummy bears, while remaining nutritious and even healthy.

Assemblers—A World Out of Dust

Let us now look at Feynman's ultimate vision: machines that assemble any matter from atomic ingredients just as kids build structures from Lego bricks. A handful of dirt contains all the necessary atoms to allow such "assemblers" to build everything we desire seemingly out of nowhere in a kind of atomic 3D printer. "Nano-3D" could soon become a new tech buzzword.

Such machines would not in fact be entirely new! They have existed on our planet for 1.5 billion years. In the two hundred different cell types of our body, nanomachines use certain building blocks (sugar molecules, amino acids, fats, trace elements, vitamins, etc.) to assemble proteins, cell walls, nerve fibres, muscle fibres, and even bone, molecule by molecule. Very specific proteins play an important role here. These are the enzymes. The energy needed for these processes is extracted from the food we eat. The whole thing is like a mini assembly line: biological nanomachines transport, produce, and process everything we need to live in various metabolic processes.

> The fact that assemblers are possible was already proven a long time ago by nature's invention of cell metabolism in living systems. As nanomachines, enzymes are the true masters of efficiency.

What prevents us humans from developing such technologies ourselves? We can even go one step further: if nanomachines can do all sorts of things, what keeps them from building themselves? This is another thing that nature has proven possible on the nanoscale: DNA and RNA are nothing but highly efficient, self-replicating nanomachines. The step towards self-replication of man-made nanomachines may not be as far away as it seems.

> Even the problem of self-replication of nanomachines has long been solved by nature: DNA can be understood as a self-replicating nanomachine.

Nanotechnology creates tremendous opportunities to improve our lives. Nevertheless, to most people the prefix "nano" is a source of great discomfort, as indeed are the terms "gene" and "atomic" which also refer to the inconceivably small. All three of these, nanoparticles, genes, and atoms, are things we cannot see directly, while the technologies that rely on them are doing more and more to shape our daily lives.

But what happens when artificial nanomachines develop a momentum of their own and become able to multiply uncontrollably and exponentially? Or when nanomaterials prove to be toxic? The first such problems have already arisen: the nanoparticles used in many products, for example in cosmetics, can accumulate in unintended places—for example in the human lung or in marine fish. What effects do they have there? Which substances react chemically with them and which can bond with their highly active surfaces? Certain studies show that some nanomaterials are indeed toxic to microorganisms.

Information and education are required to better assess not only the opportunities, but also the effects of nanotechnologies. This will apply in particular to the technology discussed in the next chapter: the quantum computer.

4

Incredibly Fast: From the Digital to the Quantum Computer

Every year, we double the amount of data produced worldwide. In 2018, as many giga-, tera-, peta-, and exabytes are produced, processed and collected worldwide as in the whole of human history prior to 2018, as data and its collection and transfer move beyond fixed computers. Smart Phones, Smart Homes, Smart Clothes, Smart Factories, Smart Cities … many "smart" things are getting connected via the Internet. And they are producing more and more of their own data.

Correspondingly, the demand for the performance of computer chips is also growing exponentially. And in fact, their computing power has approximately doubled every 18 months over the last 50 years. The growth in the number of components per unit area on integrated circuits follows a law formulated in 1965 by the future co-founder of Intel, Gordon Moore. (The fact that the total amount of data grows even faster than the performance of individual computers is due to the fact that the number of data-producing devices is increasing equally fast.)

Concerns that the "Moore's Law" would at some point lose its validity go back 25 years. The reason is that the increasing miniaturization of components leads to two problems:

- Electrons moving through ever smaller and more numerous circuits cause the chips to heat up more and more.
- But there is a more fundamental second problem: electronic structures have now become less than 10 nm in size. This corresponds to about 40 atoms. In transistors of this size, the laws of quantum physics prevail, and this causes the behaviour of the electrons to become hopelessly unreliable.

© Springer Nature Switzerland AG 2018

L. Jaeger, *The Second Quantum Revolution*, https://doi.org/10.1007/978-3-319-98824-5_4

In 2007 Moore himself predicted the end of his law; he gave it another 10 to 15 years back then. And indeed, for the first time ever, the 2016 roadmap presented by the semiconductor industry for the following year's chip development no longer followed Moore's law.

Thanks to the creativity of nano-engineers, however, it is likely that it will still be possible to construct even smaller and faster electronic structures, thereby postponing the end of "classical" miniaturization for a few more years. But what then? How long can we rely on the possibilities for merely increasing computer chip performance?

> The fact that Moore's Law will lose its validity does not mean that we have reached the end of the flagpole with regard to further increasing the efficiency of information processing.

However, there is yet another way to build much faster computers, even billions and billion of times more powerful: *quantum computers*. Such computers work in a completely different way to conventional computers. Instead of suppressing the quantum properties of the electrons and the problems associated with the ever-increasing miniaturization of the components, a quantum computer explicitly exploits these properties in the way it processes data. With the help of such machines, we could solve problems that are far too complex for the "supercomputers" used today in physics, biology, weather research, and elsewhere. The development of quantum computers could trigger a technological revolution that would shape the 21st century to the same extent that the development of digital circuits shaped the 20th century.

> Quantum computers should enable computing speeds beyond our imagination.

Today's Computer—A Concept from the 1940s

Although the miniaturization of computer chips has forced computer engineers to consider quantum mechanical laws, the principles and functioning of today's computers are still entirely based on classical physics.

Thus the first computers from the 1940s consisted of tubes and capacitors, and the transistor, originally a "classical" component, is still a centrepiece in any computer today. "Transistor" is a short for *transfer resistor*, which just means that an electrical resistance is controlled by an electrical voltage or current. The first patent for a transistor was filed in 1925. Shortly afterwards, in the 1930s, it became clear that elementary arithmetical operations can be carried out by the targeted control of the electric current (for example, in diodes). There are two main reasons why point contact transistors, triodes, and diodes based on electron tubes can only be seen in technology museums today: lack of computing speed and energy consumption.

The components have changed, but the basic operation of today's computers is still based on the architecture formulated by the Hungarian mathematician and physicist John von Neumann in 1945. The core of von Neumann's reference model for a computer is the memory card, which contains both program instructions and (temporarily) the data to be processed. The data is processed sequentially, i.e., step by step, in single binary computation steps, managed by a so-called *control unit*. Computer scientists speak of a "SISD architecture" (*Single Instruction, Single Data*).

> Although transistors and electron tubes have been replaced by smaller and faster field effect transistors on semiconductor chips, the architecture of today's computers has remained the same since it was first formulated.

How does this sequential processing of information in computers work? The British mathematician Alan Turing described the basic data blocks and their processing theoretically in 1936. The most elementary information units therein are the *binary digital units*, or "bits". Binary means "two-valued", because a bit can assume either the state "1" or the state "0", like a light switch which is either on or off. The term "digital" derives from the Latin *digitus*, or finger, from back in the days when people counted with their fingers, while "digital" today means that information can be represented by numbers.

Electronic data processing in today's computers consists in converting incoming information in the form of many consecutively arranged bits into an output likewise in the form of many consecutively arranged bits. As with the production of chocolate bars on the assembly line, blocks of individual bits are thereby processed one after the other; for a letter, for example, a block of eight bits, a so-called "byte", is required. For single bits, there are only two processing possibilities: a 0 (or 1) remains a 0 (or 1), or a 0 (or 1) changes to a 1 (or 0). The basic electronic components of digital computers, known as logic gates,[1] are therefore always the same simple basic electronic circuits, realized by physical components such as transistors, through which information is passed in the form of electric signals. Several such gates are then connected, thereby enabling more complex operations such as the addition of two numbers.

> Every computer today is a Turing machine: it does nothing else but sequentially process information encoded by zeros and ones, transforming them into an output also encoded by zeros and ones.

However, this simplicity in the information processing has a price: a huge number of zeros and ones need to be processed to handle the amount of data required in today's complex computer applications. The computing power of a computer increases linearly with the number of available computational blocks. With twice as many circuits, a chip can process information twice as fast. Today's computer chips operate in the gigahertz range, i.e., a billion operations per second. This requires billions of transistors. To be able to pack this many transistors on chips the size of a thumb nail, the circuits must be microscopic. Only then can the overall size and energy requirements of such fast-switching systems be kept under control.

Essential for the miniaturization of the elementary computing units on integrated circuits in microchips was the transition from the electron tube to the semiconductor-based bipolar or field effect transistors, which were invented in 1947. These microscopic transistors are built on doped semiconductor layers. This is where quantum physics comes into play. In order to understand and control what is happening in these semiconductors, we require a quantum mechanical model for the motion of the electrons inside them (this is the so-called "band model" of electronic energy states in metallic conductors).

> An *understanding* of quantum physics was not necessary for the digital revolution of the 20th century, but a prerequisite for the extreme miniaturization of integrated circuits was the ability to *apply* it.

---

[1] The basic logic gates consist of the operations AND, OR, NOT, NAND, NOR, EXOR, and EXNOR.

A New Way to Calculate

In his 1981 lecture *Simulating Physics with Computer*, as part of a conceptual reflection on quantum theory, Richard Feynman raised the question of whether the quantum world could be simulated by means of a conventional computer.[2] The problem here comes from the probabilities associated with quantum states, because quantum variables do not assume fixed values. Indeed, at any given time, they fill an entire mathematical space of possible states. This increases the scope of the calculations exponentially. Sooner or later, any conventional computer will be overwhelmed, Feynman concluded.

But he then asked whether this problem could be solved with a computer that itself calculates only with state probabilities, in other words, a computer whose internal states are themselves quantum variables. Such a quantum computer would in turn explicitly exploit the bizarre quantum properties of atomic and subatomic particles. Above all, it would possess a fundamentally different structure and functionality than the von Neumann architecture of today's computers. Instead of processing sequentially bit by bit like a Turing machine, it would calculate in parallel on the numerous states assumed simultaneously by the quantum variables. The elementary information units in a quantum computer are therefore no longer "bits", but "quantum bits", or "qubits" for short. Unfortunately, this name is misleading, because it still contains the word binary, and that is exactly what quantum bits are no longer.[3] The nature of information in qubits is very different from conventional information.

> Quantum bits, or qubits, are no longer either 1 or 0, but can accept both states simultaneously, as well as all values in between. A qubit can therefore contain much more information than just 0 or 1

This particular capacity of qubits is due to two bizarre properties that are only possible in the world of quantum physics:

1. Superposition: Quantum states can exist in superpositions of classically exclusive states. In the micro world, the light switch can be both on and

---

[2]Published in R. Feynman, International Journal of Theoretical Physics, Vol. 21, Nos. 6/7 (1982).

[3]The name "qubits" goes back the American theoretical physicist Benjamin Schumacher, see: B. Schumacher, *Quantum coding*, Physical Review A 51 (4): 2738–2747 (1995).

off at the same time. This makes it possible for a qubit to take the states 0 and 1 simultaneously—and also all states between 0 and 1.

2. Entanglement: Several qubits can be brought into entangled states, in which, as if coupled by an invisible spring, they are linked in a non-separable whole. Through a "spooky action at a distance"—a term that Albert Einstein invented in irony to express his disbelief regarding this quantum phenomenon—they are in some kind of direct contact with each other, even when they are spatially well separated from each other. It is as if each quantum bit knows what the others are doing and is directly affected by their behaviour.

> Superpositions and entanglement once led to heated discussions among the fathers of quantum physics. Now they have become the foundation of an entirely new computer architecture.

Due to the completely different nature of qubits, calculations on a quantum computer differ fundamentally from those on a classical computer. Unlike a conventional logic gate, a *quantum gate* (or quantum logical gate) is not a technical building block that converts individual *bits* into one another in a well-defined way, but rather describes an elementary physical manipulation of one or more (entangled) *qubits*. Mathematically, a given quantum gate can be described by a corresponding (unitary) matrix that operates on the states of the qubit ensemble (the quantum register). How exactly such an operation and the flow of information will look in each case depends on the physical nature of the qubits. For the concrete technical realization of quantum gates remains open.

Exponential Computing Power

Not much can be done with a single qubit. Only the entanglement of many qubits, which are combined in quantum registers, allows for the high-level parallelization of the operations that make quantum computers so powerful. It is as if many chocolate factories started their production lines at the same time. The more qubits you have, the more states you can process in parallel. Unlike conventional computers, whose computing power increases *linearly* with the number of computational components, the computing power of a quantum computer increases *exponentially* with the number of qubits used.

The performance of a quantum computer is not doubled when 100 more qubits are added to 100 qubits. In principle, it is already doubled when only a single qubit is added to the 100 qubits.

In theory, if we add 10 qubits, its performance increases by a factor of 1000, although in practice other effects play a role that limit the increase (see below), with 20 new qubits the quantum computer is already a million times faster, with 50 new qubits a million billion times faster. And with 100 new information carriers, when the performance of a conventional computer has just doubled, the increase in the performance of a quantum computer can hardly be expressed in numbers any longer.

Even quantum computers with only a few dozen qubits have an incomparably higher computational power than common computers.[4]

We should note at this point that the massive parallelization by entangled states is not quite comparable to the way parallel assembly lines work in chocolate factories. The way information is stored and processed in entangled quantum systems is very different from the way ordinary digital computers work with information. Quantum computers do not literally work in parallel, they rather organize the information in a way that it is distributed over many entangled components of the system as a whole, and then process it in a very oddly parallel way.

The following example will illustrate this.[5] For an ordinary classical 100-page book, with each page the reader reads, he/she acquires a further 1% of the book content. Once all the pages have been read, the reader knows everything in the book. In a fictitious quantum book, in which the pages are entangled, things are different. Looking at the pages one at a time, the reader sees only random gibberish, and after reading all the pages one after the other, he/she still knows very little about the content of the book. Anyone who wants to know its content must look at all its pages at the same

---

[4]However, so far there are only a few known algorithms that could really use the exponential computing power of a quantum computer. Such algorithms are very hard to find. So not just any arithmetic operation can be made exponentially faster with the use of a quantum computer.

[5]This illustration is taken from J. Preskill, *Quantum Computing in the NISQ era and beyond*, https://arxiv.org/pdf/1801.00862.pdf and is based on a keynote speech at the conference *Quantum Computing for Business*, 5 December 2017 (video on www.q2b.us).

time. This is because in a quantum book the information is not printed on the individual pages, but is encoded almost exclusively in the correlations between the pages.

The concept of qubits and quantum computers remains largely theoretical for the moment. However, in recent years quantum engineers have made some remarkable progress in their endeavour to get quantum computers to work in practice. There exist many different approaches to construct qubits and entangle them. In principle, the aim is always to capture individual quantum systems, such as atoms or electrons, using some clever tricks, entangle them, and then manipulate them accordingly[6]:

- One attempt is to fixate ions (electrically charged atoms) by means of electric and magnetic fields, and let them oscillate in a controlled manner, thus linking them together as qubits.
- Another approach works via the coupling of atomic spins, which are aligned by external magnetic fields in the same was as in nuclear magnetic resonance technologies.
- Qubits can also be created with the help of so-called quantum dots. These are special places in a solid where the mobility of electrons is severely limited in every direction. According to the laws of quantum physics, this means that energy can no longer be emitted continuously, but only in discrete values. These points therefore behave like huge artificial atoms.
- Other research groups seek to realize quantum computers by injecting electrons into loops in circular superconductors (called *superconducting quantum interference devices* or SQUIDs), these being interrupted by very thin layers of insulator. Here lies a special focus of companies such as Google, Microsoft, IBM, and Intel. The research exploits the Josephson effect, according to which the Cooper electron pairs of the superconductor (see Chap. 9) can tunnel through the insulator barrier. They can thereby be in two different quantum states—flowing both clockwise and anticlockwise at the same time. Such superpositions can be used as qubits and they can be entangled.
- Special chemical compounds could also be suitable as qubits. One example is provided by a complex of a vanadium ion which is enclosed by organic sulfur compounds. The shell shields the spin of the ion inside so well that its state (and possible entanglements) are preserved for a long time.

---

[6]Different practical ways to implement quantum computers are well described in: S. Aaronson, *Quantum Computing since Democritus*, Cambridge (2013).

- A still purely theoretical concept is the so-called topological quantum computer. The idea behind it originally comes from mathematics, and it is not yet clear if and how it can be implemented physically. It is based on so-called anyons (not to be confused with the anions from aqueous solutions). These are states in two-dimensional space that display particle properties. They are thus also referred to as "quasi-particles". Anyons, for example, occur at insulator interfaces. Such topological qubits should form relatively stable networks and would be far better insulated from disturbances than qubits in other concepts.

> Many scientific groups around the world are working to build a quantum computer. The suspense is rising! Which approach will prevail?

Problems Quantum Computers Can Solve—And New Ones They Could Create

Five problems in which today's computers soon reach their limits, no matter how large they are, reveal the potential of quantum computers:

1. Cryptography: Almost all common encryption methods are based on factorizing the product of two very large primes. In order to decrypt the message, one must determine which two primes a given number is composed of. For the number 39 this is straightforward: the associated primes are 3 and 13. But beyond a certain size of number, this task can no longer be solved by a classical computer. In 1994 the computer scientist Peter Shor developed an algorithm for use with a quantum computer which could factorize the products of very large prime numbers into their divisors within minutes.[7]

2. Solution of complex optimization tasks: Mathematicians consider the task of finding the optimal solution among many variants particularly tricky. The standard problem is that of the traveling salesman. The task is to choose the order in which he should visit several locations so that

---

[7]P. Shor, *Algorithms for quantum computation: Discrete logarithms and factoring*, Proc. 35nd Annual Symposium on Foundations of Computer Science, IEEE Computer Society Press (1994). Shor was able to show that while the runtime of classical factorization algorithms increases exponentially with the size of the prime number, the increase is only polynomial with the size of the number with his algorithm running on quantum computer. The basic building block of his method is a complex mathematical operation, the so-called *Quantum Fourier Transformation*.

the entire travel route is as short as possible. With only 15 cities, there are over 43 billion combinations of paths possible, with 18 cities that number increases to more than 177 trillion. Related problems occur in industrial logistics, in the design of microchips, or in the optimization of traffic flows. Even with a small number of points, classical computers already fail to determine the optimal solutions in any reasonable time. Quantum computers are expected to be able to solve such optimization problems much more efficiently.

3. A significant application could lie in the field of artificial intelligence: The deep neural networks employed in this field come with combinatorial optimization problems that quantum computers can solve much better and faster than any classical computer. In particular, quantum computers could detect structures much faster in very noisy data (highly relevant in practical applications), and accordingly learn much more quickly. Thus the new "mega buzzword" *quantum machine learning* is currently making the round, combining two buzz words that already individually excite the imagination of many people.

4. Searches in large databases: When searching unsorted data sets, a classical computer is forced to examine each data point individually. The search time therefore increases linearly with the number of data points. For large amounts of data, the number of computation steps required for this task is too large for a classic computer to be practical. In 1996, the Indian–American computer scientist Lov Grover published a quantum computer algorithm for which the number of computational steps needed increases only as the square root of the number of data points. Instead of taking a thousand times as long for a billion data entries compared to a million data points, the task would only take a little over 30 times as long with a quantum computer and the Grove algorithm.

5. Use in theoretical chemistry: Quantum computers could massively improve models of electron behaviour in solids and molecules, especially when entanglement itself plays a major role in that behaviour. For as we know today, the calculation and simulation of quantum systems involving interacting electrons is actually best done using computers that themselves have quantum mechanical properties, as Feynman had already observed in 1981. Today, theoretical physicists and chemists often deal with complex optimization problems that involve choosing from many alternatives the best possible, i.e., energetically most favourable configuration of electrons in an atom, molecule, or solid. For decades, they have

been dealing with such problems with rather limited success.[8] Quantum computers could directly map and model the quantum behaviour of the electrons involved instead of applying algorithms to qubits, because they behave like a quantum system themselves, while a classical computer often needs to pass by a crude simplification of such systems.[9] Physicists therefore speak of *quantum simulators*. Alán Aspuru-Guzik, a pioneer in the simulation of molecules on quantum computers, says: "Right now we have to calibrate constantly with experimental data. Some of that will go away if we have a quantum computer."[10]

Of course, the applications of quantum computers are of great interest to government institutions, too. For example, intelligence agencies could gain access to sensitive data of other governments (or their citizens) by using a quantum computer and its code-cracking abilities. Edward Snowden has made it public that the American NSA has shown significant interest in the technology. In the same way, quantum computers could open up a new era in industrial espionage, as data from companies would no longer be entirely safe.

Some physicists even hope to be able to use quantum computer to calculate *all* the problems in nature which are difficult to calculate on classical computers because of their complex quantum properties. Specifically, quantum computers could help to do the following:

- Calculate the ground and excited states, but also the reaction dynamics in complex chemical and biological molecules. This would be important, for example, for the development of active pharmaceutical ingredients, for building even more functional catalysts, or for optimizing the Haber–Bosch process for producing fertilizers.
- Elucidate the electronic structures in crystals, which would significantly advance solid state physics and materials science. New findings in these fields would give nanotechnology a tremendous boost. One example is the precise calculation of the properties of potential new energy storage devices or components in molecular electronics. Another application of

[8]R. Laughlin, *A Different Universe: Reinventing Physics from the Bottom Down*, New York (2006).

[9]The much cited "quantum computer" of the firm D-Wave is a special form of a quantum simulator.

[10]https://www.technologyreview.com/s/603794/chemists-are-first-in-line-for-quantum-computings-benefits/.

the utmost importance would be the search for new high-temperature superconductors.

- Calculate the behaviour of black holes, the evolution of the very early universe, and the dynamics of collisions between high energy elementary particles.

If scientists could better predict and understand molecules and the details of chemical reactions than they do today with the help of a quantum computer, they might find new types of medication on a weekly basis, or develop much better battery technologies within a month.

> Quantum computers are a threat to global data security. At the same time, they could enable scientists to solve previously unsolvable problems in many scientific disciplines and thus make tremendous advances in technological innovation.

When Will the Quantum Computer Come?

In the spring of 2016, the IT company IBM announced that it will provide public access to its quantum computing technology in the form of a cloud service. Interested parties could log into a 5-qubit quantum computer via the Internet as part of the *IBM Quantum Experience* and create and execute programs using the provided programming and user interface. IBM's goal was to accelerate the development of larger quantum computers. In January 2018, the firm offered selected companies access to the 20-qubit versions of their quantum computer. And apparently, prototypes with 50 qubits already exist.

Then in the summer of 2016, the company Google announced that a 50 qubit quantum computer would be available by 2020. They later brought this timeline forward to 2017 or early 2018. In March 2018, around the date of completion of the manuscript of this book, Google announced the introduction of a new 72-qubit quantum processor called *Bristlecone*. According to IBM, there will be quantum processors available by the mid to late 2020s consisting of up to 100 qubits. According to most quantum engineers a quantum computer with about 50 qubits could exceed the computing capacity of any supercomputer today—at least for some important computational problems. Google speaks in this context of *quantum supremacy*.

We will know very soon what new possibilities arise with real quantum computers. We could witness the beginning of a new era.

However, some difficult problems remain to be solved along the way to building functioning quantum computers. The most significant of them is that entangled quantum states decay very fast under the ubiquitous influence of heat and radiation—often too fast to perform the desired operations without error. Physicists speak in this context of the "decoherence" of the quantum states. This phenomenon will be discussed in detail in Chap. 26.

Working with qubits seems almost like writing on the surface of water, rather than on a piece of paper. The latter can last for centuries, while any writing on water disappears within a fraction of a second. So it is important to be able to work with quite crazy speeds (and by the way, even the speeds at which classical computers process data are hard for us humans to imagine).

To overcome this hurdle, quantum engineers are pursuing a twofold strategy. On the one hand, they are trying to extend the lifetime of the qubits, and hence reduce their susceptibility to errors, and on the other hand they are developing special algorithms to correct the errors that occur (this is called quantum error correction). Physicists are able to limit the effects of decoherence with the help of ultra-cold refrigerators. In addition, the techniques for handling decoherence-related errors in individual qubits are becoming better and better today. There is thus hope that the reliability of quantum computers will increase significantly in the future. So far (as of spring 2018), however, the efforts of the quantum engineers have not yet yielded reliably functioning quantum computers.

Companies like IBM, Google, Intel, Microsoft, and Alibaba are working on making quantum computers a reality in the next few years. They claim to have made some significant progress in the recent past.

The Quantum Internet

Due to the sensitive nature of the qubit, the transport of qubit information is technically much more involved than the transport of electrons in classical computers (as it happens in any electric cable) or of electromagnetic waves in the global internet. Nevertheless, quantum information can already be transmitted via optical fibre over hundreds of kilometres with little loss of information. This is possible due to entanglement. Physicists speak of *quantum teleportation* in this context. The name is somewhat unfortunate because quantum teleportation has nothing to do with the transport *of matter* between two points without traversing the space between these points, as described in the popular science-fiction literature. Quantum teleportation rather involves the transfer of quantum properties of particles, i.e., quantum *states* (qubits), from one place to another. Thus, only *quantum information* is transmitted, but in such a way that there is no transmission path along which the information passes from sender to receiver.

In theory, entangled particles can be arbitrarily far apart, without the entanglement between them ever dissolving. Physicists have suggested since the 1990s that this property makes quantum teleportation possible in practice. The basis of this technology is that two quantum particles (for example, photons) are entangled in a common quantum physical state and then spatially separated, without destroying their common state. One of the particles is sent to the receiver, the other remains at the sender.

So much for the preparation. Now the actual information transfer can begin. At the sender, a simultaneous measurement of the entangled qubit and the qubit to be teleported is performed (a so-called "Bell measurement").

According to the laws of quantum physics, the measurement of the sender's particle automatically and instantaneously determines the state of the entangled particle at the receiver, without any direct interaction taking place between them.

The result of the measurement at the sender is then transferred via a conventional communication channel to the receiver. With the measurement, the receiver qubit together with the entangled qubit at the receiver is projected onto one of the four possible states they can jointly be in. Using the information about the result of the measurement at the sender, the receiver qubit can be transformed to be in the same state as the sender qubit. In this way, the desired (quantum) information is brought from the sender to the receiver without physically transporting a particle. (Of course, the receiver can equally well become the sender by manipulating his/her particle in the same way.)

Because the result of the measurement is transmitted conventionally, i.e., not instantaneously, quantum teleportation is not about transporting information faster than light, but rather about transferring quantum states reliably from one place to another.

> Quantum teleportation opens up the possibility of transmitting, storing, and processing qubits, i.e., quantum information. Thus, in addition to the quantum computer, a quantum internet appears to be within reach.

Quantum technologies will soon dramatically change our world. But to fully appreciate them, we need to take a step back and understand how physicists have learned to describe the world of atoms. For this purpose we will delve more deeply into the bizarre world of quantum physics in the next part of the book.

Part II

Quantum Worlds—The Bizarre in the Very Small

5

Contradictory Atoms: Philosophical Problems with the Smallest Building Blocks of Nature

To see if, through Spirit powers and lips,
I might have all secrets at my fingertips.
(…) That I may understand whatever
Binds the world's innermost core together.

At the beginning of his famous play *Faust*, Goethe expresses what drives his protagonist: it is the search for knowledge. The question of what holds the world together at its innermost core had already been posed by the pre-Socratic philosophers 2,500 years ago. They were the first to doubt that the events in nature can be explained by the will of a few gods. This constituted the beginning of (Western) philosophical thought, which no longer interprets world events as the playthings of supernatural powers and divine interests, but seeks rational explanations as to why the world is the way it is.

Of central importance to the early Greek philosophers was the question of how change can come about. How does a seed become a colourful flower? How can an egg turn into a chicken? Everywhere around us we see change. The question raised by the pre-Socratics was this: if something changes, does that something not always change in relation to something else that itself remains unchanged? You only notice that a river flows when you have the bank in view. It is exactly this immutable feature that should "hold the world together at its core", and that constitutes the foundation of all things.

The crucial question was—and still is—therefore: is there a fundamental substance in the world that underlies all observable changes, that is itself unchanging, self-existent, and independent of everything else?

© Springer Nature Switzerland AG 2018
L. Jaeger, *The Second Quantum Revolution*, https://doi.org/10.1007/978-3-319-98824-5_5

In order to be able to answer the question about the foundation of all things, all happenings, and all changes, the Greek philosophers searched for the absolute and ever changeless components of nature.

The Greek philosopher Leucippus and his disciple Democritus developed the idea that all things are always made of the same indivisible and unchanging particles. Any observable change in nature, they declared, was a change in the *composition* of these elementary particles, while they themselves remained the same. They gave these particles a name that would stick right through to modern times: *atom* (the Greek *a-tom* means "indivisible"). To explain the diversity of nature—rocks, plants, the human body, and so on—there had to be many different kinds of atoms. Democritus portrayed them descriptively with hooks and loops, some smooth and round, others angular.

Of course, the atomic theory of Leucippus and Democritus was of a purely speculative nature. The pre-Socratics had no particle detectors or radioactive sources at their disposal, so could not defend their theories like today's physicists.

The idea of atoms has been familiar to humans for 2,500 years. But their existence remained unproven until fairly recently.

Philosophical Contradictions

The train of thought which led Leucippus and Democritus to the idea of atoms, came about by considering the philosophical treatment of the continuum and therefore concerned the infinitely small. Some pre-Socratics, in particular Leucippus' teacher, Zeno of Elea,[1] had devoted much thought to this question. They were concerned with the problem of imagining infinite repetitions of the smallest processes, or infinite sums of smaller and smaller units. With questions about the nature of space, time, and movement, Zeno had stumbled upon a few paradoxes, the most famous of which concerns a race between fast-paced Achilles and a turtle:

---

[1]It is unclear whether Zeno really was Leucippus' teacher. According to other sources, Parmenides was the teacher of both Leucippus' and Zeno.

In a race with Achilles a turtle gets a head start. Can Achilles ever overtake the turtle? As soon as he arrives at the turtle's starting point, the turtle will already have moved a little further. When Achilles comes to this point, the turtle will already have moved a little further on, if only by an even shorter distance. And so on and so forth.

If there are infinitely many infinitely small units of measure, then Achilles will never overtake the turtle, according to Zeno's argument.

Here was an irreconcilable contradiction for Zeno: *logically*, Achilles can never reach the turtle, even if his distance from it is getting smaller and smaller; but *experience* tells us that Achilles will quickly overtake the turtle. Zeno's way out of this paradox (and that of his teacher Parmenides) was to conclude that our everyday perception of diversity and movement is a mere illusion. In reality, neither exist.[2]

Leucippus gave a very different answer than his teacher: maybe there is no arbitrarily small unit of measure in nature! Leucippus then transferred this idea to matter. The following thought experiment was supposed to support his assumption. The constituent parts of any matter must be separated by spaces, along which one can divide them. If matter could be divided again and again infinitely many times, then it would ultimately consist only of these "empty" dividing lines, that is of nothing. This is in contradiction with our experience that matter exists in the first place.[3] Leucippus concluded that matter *must* consist of smallest, indivisible particles in which there are no more dividing lines.

Another approach, which comes to the same conclusion is as follows. Matter must be assembled from the smallest particles, because only these could provide solidity. If there were no smallest building blocks, all matter would dissolve like water.

The hypothesis that the world is made up of atoms seems to be a logical imperative. However, it also inevitably leads to contradictions.

---

[2]The reflections of Zeno and his contemporaries on the summation of infinitely many terms of infinitesimal sizes has long since become obsolete. Modern mathematics has no difficulty showing that such infinite series can converge to finite quantities.

[3]Interestingly, physicists know today that much of the mass of an atom consists in the binding energy of the quarks in its nucleus, and not the "bare" quark masses themselves.

Doesn't such an atom, however small it may be, have to fill a certain space? It cannot be *infinitely small*, because only the spatiality of the atoms makes spatiality of matter possible. But if an atom occupies a certain space, smaller spaces than the one it occupies are conceivable. And as soon as we can imagine parts of atoms, they can no longer be thought as indivisible. Thus there cannot be indivisible atoms. (In fact, atoms have turned out to be divisible—how else could nuclear fission take place? Equally, protons, neutrons, electrons, photons, etc., if they take up any space at all, cannot be considered as "smallest particles". So high school physics is not as viable as we might think.)

In his major work *The Critique of Pure Reason*, the 18th century philosopher Immanuel Kant brought this contradiction concerning the existence and properties of atoms to a head. In what he calls the "antinomies of pure reason," Kant describes fundamental contradictions in our thinking. The second of Kant's four antinomies deals with the question about the smallest indivisible particles of matter. Kant provides logical evidence for both hypotheses, i.e., that there exist such indivisible atoms and also that they cannot exist.

His explanation was that only our (transcendental) forms of intuition and categories of reason make our experiences possible—any conclusions from our reason are valid only for what can be experienced. Because we cannot have any direct experience of atoms, we inevitably fall into contradictions when we try to think about them. We will meet Kant in more detail when it comes to the philosophical interpretation of quantum theory in Chap. 16.

> Philosophers came to the conclusion that the existence of smallest indivisible particles can be perfectly well proved logically by the faculty of reason. But their non-existence too!

The Cave Allegory Versus Atoms with Hooks and Lugs

The successors of Democritus, including the Greek philosopher Epicure and some of his followers, like Lucretius, used Democritus' theory to develop a consistent materialist philosophy, according to which even the soul was composed of smallest (soul) particles. But later, the idea of atoms was consigned to oblivion for a very long time.

For us humans today, it seems clear that the world is made up of atoms, the smallest material building blocks. The fact that the early atomic theories disappeared for so long does not mean that people had no time to do philosophy. On the contrary, the search for "what holds the world together" has never long been set aside. But if not atoms, what else did the philosophers think could be the basis of the world?

Plato and Aristotle, the most important philosophers of antiquity (and through their teachings also the most famous of the Middle Ages) rejected the materialistic atomic doctrine of Democritus altogether. In their view, the true and ultimate causes of world events were not material, but rather spiritual and thus supernatural. In his *theory of ideas*, Plato, a younger contemporary of Democritus, emphasized the transcendent as the origin and principle of all being. He denied that the observable things in nature (the "phenomena") had any universal qualities. His approach was rather to say that our concrete everyday perceptions are just imperfect images of perfect (spiritual) ideas. Only these possess universally valid properties of the kind worth thinking about.

In his *Politeia*, Plato described his thoughts in the most famous parable of ancient philosophy: the cave allegory. Human beings are like the denizens of a cave who merely see the shadow of the things outside projected on the cave wall through the entrance hole. The things we perceive as real in our everyday lives are essentially just pictures of something beyond the sphere of our experience, i.e., spiritual ideas.

> From Plato to the early modern period (about 1500), most philosophers assumed that the basis of all being in the world was not matter but spiritual principles.

Of course, philosophers also recognized elementary structures in the world of ideas. For Plato, the order and structure of the material world resulted from mathematics. He saw the basic principle of everything material in the five regular convex polyhedra, today called "Platonic solids."[4] Because they offer the highest degree of symmetry and beauty, Plato was convinced that a creator of the world must inevitably have used these spiritual forms as a template for every material structure.

---

[4]These are the tetrahedron (bounded by four equilateral triangles), the hexahedron (the cube bounded by six squares), the octahedron (bounded by eight equilateral triangles), the dodecahedron (bounded by 12 equilateral pentagons), and the icosahedron (bounded by 20 equilateral triangles).

His pupil and successor Aristotle believed that all worldly matter was composed of four elements: earth, fire, water, and air. To this he added a fifth element called "quintessence" that made up the heavenly bodies. But for him, too, matter was not the basis of the world. According to Aristotle, an everlasting mover (God) was the immutable entity that held the world together.

Plato and Aristotle had already more or less completely ousted the materialistic world view of Leucippus and Democritus. The rest was done by Christian religious censorship. The philosophers of the European Middle Ages referred directly to Plato and Aristotle and consistently banished any atomistic thinking, in which they saw the mortal enemy of true philosophy and theology. For with it disappeared any need for God's existence. The followers of Epicurus had already been persecuted in early Christianity, and in the late Medieval Ages his teachings had almost been eradicated.

> For 2,500 years, Democritus and Epicure's idea of a world made up of tiny particles was ridiculed and its advocates persecuted. Spiritual and transcendental principles were supposed to determine the material world.

The Path Towards a Physical Theory of Atoms

It was not until two and a half millennia after Democritus that natural scientists allowed atoms to take back their rightful place in the world. With the beginning of the Scientific Revolution in the 17th century, concrete observations and experiments replaced the purely theoretical discussions previously put forward to explain nature.

Physicists observed that gases (for example, air) could be compressed in a tightly closed container. This implied that there was ample free space in a gas, which could be reduced by applying pressure. We can now imagine that a gas consists of free-flying particles that whirl wildly around in their container like little balls. To their astonishment, scientists found a universal relationship between volume, pressure, and temperature that applies to all gases. In 1811, the Italian Amadeo Avogadro concluded that, at constant temperature and pressure, the same gas volumes always contain the same number of particles. Since the laws were the same for all gases, the smallest particles making them up had to possess universal properties. This conjecture was first formulated by the Englishman John Dalton in 1808 in his work *A New System of Chemical Philosophy*.

By observing gas temperatures in confined spaces when gases are exposed to different pressures and temperatures, physicists concluded that gases consist of tiny particles.

Dalton wondered whether *all matter*, i.e., not just gases, but also liquids and solids, might be composed of these smallest particles. There already existed some hints in favour of this hypothesis. For example, chemists had observed that chemical substances, which arise from combinations of other substances, always emerged from integer ratios of their starting materials. This law could easily be explained by the fact that substances were composed of smallest elementary particles which always combined in the same proportions. For example, hydrogen and oxygen always turn into water in a ratio of 2:1. By contrast, ratios of, for example, 1 to 1,735 or 2,834 to 4,925 never occur.

In order to explain the great diversity of chemical substances, Dalton had to assume that there were many types of atoms. He created the idea of chemical elements. However, unlike Democritus, the different atoms should not differ in colour, feel, and shape, but only in their specific atomic weight. To the atom of hydrogen he assigned a weight of one unit, while he determined the atoms of all other elements as integral multiples of that.

As basic building blocks of matter, atoms always come in integer numbers. These numbers can be 27, 52, or 2,189,983, but never 1.64. In other words, they are quanta. We can say that Dalton's and Democritus' atomic theories were the first quantum theories.

Dalton's atomic theory quickly prevailed in the scientific community of the day. It was as though what had hitherto seemed an inextricable knot was finally cut: the new theory opened the way to explaining the structure of matter, picturing the complexity of chemical processes, and interpreting the diversity of material forms in nature as distinct configurations of unitary fundamental substances. Everything suddenly began to fit together.

The next big question was: what forces hold atoms together? To explain this, physicists needed a new theory that was to be born in the 19th century—the theory of electricity and magnetism. It turned out that the essential forces in the atom are of electromagnetic nature. And even for electricity, there is a smallest unit, as physicists discovered in the late 19th century:

the charge of the electron. But what atoms and electrons were exactly, they had to leave open for another few decades.

The Problematic Framework of Classical Physics

The philosophical speculations of Democritus and the "antinomies of pure reason" of Kant had shown that the idea of a smallest indivisible particle as a fundamental building block for all material existence brings with it some logical pitfalls. The new atomic theory proved to be equally contradictory. Indeed, physicists were unable to describe the smallest particles theoretically.

For example, the concept of mass points, as used in classical Newtonian mechanics, becomes questionable in the atomic world. In mechanics, when a stone is being transported up a ramp, its mass distribution can (at least conceptually) be assumed pointlike without the theory losing any accuracy. But is that also true for atoms, or for the elementary carriers of electrical charge in electrodynamics? Do they have a spatial extension? And if so, how much room do they fill out? And with what? On the other hand, if they come without spatial extension, i.e., if they are infinitely small, then they must inevitably possess an infinitely high mass or charge density. This idea is very difficult for physicists to accept.

> Attempts to describe the smallest particles in classical physics led to irresolvable logical contradictions. The world view of classical physics was therefore anything but firmly and self-consistently established.

The world of the microcosm which physicists entered in the late nineteenth century forced them into a tremendous level of conceptual and perceptual abstraction. For in their exploration of the atom during the first 25 years of the 20th century, they had to recognize that, in the realm of the very smallest, the metaphysical navigation maps of the day were no longer reliable. What was revealed was a world that differed significantly from the world of everyday experience, making a complete break with millennia-old philosophical and metaphysical notions.

On their journey into modern physics, physicists were again and again to be reminded of the philosophical contradictions in the classical concept of the atom:

- "There must be atoms as fundamental building blocks of matter" versus. "There can be no indivisible fundamental building blocks in nature".
- "Matter is the foundation of the world" versus. "Spiritual (or mathematical) principles form the foundation of the world".
- "Electric charges and massive particles have spatial extension" versus. "Electric charges and massive particles are pointlike".

For 2,500 years, atomic theories have led to unsolvable contradictions. As physicists began to develop a *physical* atomic theory, they thus had to be prepared for serious *philosophical* problems.

6

Natura facit saltus: On Quantum Jumps and Particles Being Created Out of Nothing

The aphorism *Natura non facit saltus* (Latin for "Nature makes no leaps") describes a fundamental assumption in occidental thinking: processes in nature do not happen discontinuously and abruptly, but continuously and predictably. An oak tree does not turn from a seedling into a majestic tree overnight, but needs a few decades for that.

We already recognized this principle in the philosophy of the pre-Socratics Parmenides and Zeno, and it played a fundamental role in the ideas of Aristotle and right up to modern Western philosophy and natural science. It appeared explicitly in the works of Leibniz and Newton, who based the new mathematics of calculus on it, and also held an important place in Kant's philosophy and Western thinking right up to the founders of modern biology Carl von Linnaeus and Charles Darwin.

> Nature makes no leaps—the whole of Western thinking was based on this certainty.

Quantum physics was to shake that assumption.

An Act of Despair

It all began in a back room of the physics institute of the Friedrich Wilhelms University (now Humboldt University) in Berlin. In the last few years of the 19th century, the physicist Max Planck was investigating a topic which at

© Springer Nature Switzerland AG 2018
L. Jaeger, *The Second Quantum Revolution*, https://doi.org/10.1007/978-3-319-98824-5_6

the time aroused little interest outside a small circle of specialists: how do material bodies absorb and release energy?

At that time, light bulbs and radiators were already in use, both being based on the principle that matter radiates energy in the form of light and heat. For light bulbs, it is a metal wire heated by an electric current that emits the desired light (and less desired heat). People had known for millennia that a body with a lower temperature radiates a different light than one with a higher temperature. Every good blacksmith recognizes when it is time to reheat the metal he is forging, just by looking at its colour (matt orange-red to a bright white glow). A candle flame has a yellow and a blue area in its flame, depending on the local temperature (and specific combustion process).[1] Physicists summarized all these observations by saying that, depending on the temperature of a body, its outgoing radiation has a different wavelength (which means, in the visible range, a different colour).

But what energy *exactly* do bodies of a certain temperature radiate as heat and what frequency *exactly* do they radiate as light? That was the problem Planck was struggling with. In his calculations he started from what were known as black bodies. These represent the ideal case in which all electromagnetic radiation of any wavelength that hits them is completely absorbed, whereas real bodies always reflect a part of it.

> Planck and his colleagues were struggling to find a formula describing the radiant energy of heated bodies that matched physical measurements.

As he was finally running out of ideas, Planck at some point assumed that the bodies did not emit heat and light continuously, but in little packages. This assumption contradicted classical physics, where energy radiation always has to happen continuously—just because nature makes no jumps. Planck called these energy packages "quanta" (from the Latin word *quantum* for "so much"). He called the smallest unit of this energy a "quantum of action". And to his surprise, using this quantum hypothesis, he succeeded in deriving a formula that corresponded precisely to what was observed experimentally.

---

[1]The blue part of the flame is caused by radiation transitions of certain excited molecules, while the bright part is due to glowing soot particles (whose light emission behavior corresponds to that of a black body).

Planck considered his quantum hypothesis as nothing more than a temporary stopgap or auxiliary assumption that he would later get rid of. Accordingly, he gave his quantum of action the name "h", from the German word "Hilfsgröße", meaning "auxiliary term".[2] He could never have imagined that this auxiliary term would remain forever in physics, but it has indeed kept its name and significance right up until today.

> The origin of quantum theory was a mathematical trick to derive the radiation formula for ideal black bodies.

Max Planck's formula also revealed a novel relationship between the energy and the frequency of radiation. His formula $E = h \cdot f$ is just as important as Einstein's later formula $E = mc^2$. It means that the energy (E) depends directly on the frequency (f) of the radiation, the proportionality factor being Planck's constant (h). This relationship caused physicists some headaches: energy here is a property of particles, while frequency is a feature of waves.

The Next Quantum Leap—Einstein's Light Particles

For a few years, Planck's quantum hypothesis did not cause much of a stir. Physicists saw in it what Planck saw in it: a pragmatic mathematical trick. The first interpretation that accepted Planck's quantum effect as a *physical* reality would be made by a young and at the time completely unknown 25-year-old physicist who was then working full-time as a Swiss civil servant, at the Patent Office in Bern, where he described himself as an "ink pooper". His name was Albert Einstein.

Planck had given the world the perplexing solution that bodies can only emit quantized radiation or energy. Five years later Einstein went one step

---

[2]Planck also managed to determine the value of this constant. But he was also able to determine the value of the so-called Boltzmann constant from it. This revealed a deeper connection between the theory of gases (in which the Boltzmann constant plays a central role) and microphysics. Later, Planck realized that these two constants, together with the gravitational constant, the electric field constant, and the speed of light, form a system of universal constants of nature, from which the universal units of length, mass, charge, time, and temperature can be derived. Today these constants are called "Planck units".

further: He claimed that *all* electromagnetic radiation occurs only in quanta. These portioned energy packages behave like spatially localized particles. This meant that light, so far always treated as a wave, was actually made up of particles!

Einstein called these particles *light quanta* (today they are called *photons*). Their existence fundamentally contradicted classical physics, as it had long been proven in numerous experiments that light was a wave.

But Einstein's hypothesis was equally capable of matching experimental observations:

- There was the "photoelectric effect", discovered by Heinrich Hertz in 1887. Hertz had observed that a metal irradiated with electromagnetic waves emits negatively charged particles. In a note on his experiments, Hertz wrote that his particle detector scored more hits with (higher-frequency) UV light than with (lower-frequency) visible light. But if light is a wave, the lower limit at which electromagnetic waves knock out particles from the metal should depend solely on the wave *height* (amplitude) associated with the incident radiation, since this is what determines the energy of the radiation, according to wave theory. It should not depend on its *frequency*.
- In 1902, the German physicist Philipp Lenard had discovered that the energy of the emitted electrons was actually quite independent of the intensity of the electromagnetic radiation (defined as the square of the wave amplitude). In addition, below a certain frequency, no electrons escaped, even at highest radiation intensity.

Both observations contradicted the wave nature of light. To explain these astonishing effects, Einstein proposed the following mechanism. A quantum of light entering the metal donates all or part of its energy, which according to Planck's formula $E = h \cdot f$ is directly proportional to the frequency of the radiation, to an electron in the metal. Thus, the now free electron has the energy $E = h \cdot f\text{-}P$, where P is the energy needed to knock the electron out of the metal compound (less than P, no electron is emitted).

For nine years, Einstein's hypothesis could not be confirmed experimentally: the energy of the incident radiation could be measured with sufficient accuracy, but not that of the exiting electrons. Only in 1914 was Robert Millikan able to measure the relationship predicted by Einstein and Planck precisely enough.

With the explanation of the photoelectric effect by his particle hypothesis, Einstein firmly established Planck's quanta in physics—an auxiliary construction had become a variable that had to be reckoned with.

Wave–Particle Duality

Einstein clearly recognized, of course, very well that his quantum hypothesis stood in stark contrast to the classical wave theory of light. He emphasized right at the beginning of his 1905 paper that the wave theory of light could claim very convincing experimental support. His new hypothesis did not throw an old hypothesis out, as had often happened in physics, but stood alongside it. That was a novelty.

At this point the contradictory nature of quantum physics revealed itself for the first time, and this would later become a subject of heated discussion: light seems to be both a spatially extended wave and a localized particle at the same time. But how is that possible? Einstein's ingenuity allowed him to immerse himself in completely new ways of thinking: instead of insisting on an *either/or*, he considered *one-as-well-as-the-other* to be possible.

Einstein's first work on quantum theory already identified a fundamental phenomenon of the micro world: wave–particle duality.

How did Einstein explain this dual nature of light? In the wave nature of electromagnetic radiation he saw the effect of a very large number of light quanta. As in thermodynamics, in which the temperature of a gas results from the average velocity of its particles, the wave phenomena of light can equally well be interpreted as temporal and spatial averages of its many photons. The strength of the electromagnetic field in one place at a given time thus results from the average light quantum density in its immediate vicinity at that point in time. In this way, the experimental phenomena typical of the wave nature of light emerge, including interference patterns and diffraction (in which waves overlap).

However, if one goes down to the atomic level, one has to consider the local spatio-temporal fluctuations in the electromagnetic wave fields, as

Einstein pointed out. This is analogous to the situation in a gas with only a few particles, whose properties can no longer be described from their statistical averages (the gas in interstellar space, which contains only an average of 1 hydrogen atom per cubic centimetre, has no temperature).[3]

Einstein described this as follows: when viewed at the atomic level, one cannot "operate with continuous spatial functions"; one has to consider light as "energy quanta located in spatial points, which move without being able to divide and can only be absorbed and produced as a whole".

> While one can describe non-local optical phenomena such as interference and diffraction extremely well by assuming the wave nature of light, the particle picture of light is better suited to localized phenomena such as light generation or light absorption.

However, not all problems were solved by Einstein's wave–particle duality. How can we imagine a spatially localized quantum of light? How much room does it take? If it does not fill any space and is thus located at single point, it will come with an infinitely high energy density, which is difficult to imagine physically. If, however, it does take up space (whence physicists do not have to abandon the notion of spatial continuity), how can we envision the energy distribution within that spatially extended light quantum? Einstein did not believe the photon possessed any interior structure. For him it had no spatially continuous extension. But he was unable to offer a coherent explanation of what a photon essentially is.

> Does a photon occupy any space or not? In this question we recognize the intellectual dilemma that philosophers from Democritus to Kant had already debated in their reflections on the atom.

Einstein was also unable to answer yet another important question: how can we describe the creation of a light particle in light emission or the destruction of a photon in the photoelectric effect? Does a photon emerge out of nowhere and disappear in the void? An answer to this question would not be found until 30 years later, with the development of quantum field theory.

---

[3]Einstein's interpretation later turned out to be incorrect, leading to great discussions among physicists. The special feature of the wave–particle dualism of quantum physics is that one can measure the wave nature even with single-photon sources, so the wave nature is contained in each individual particle, and is not just a collective effect (see Chap. 9).

From Thomson's Raisin Bread Model to Bohr's Quantum Leaps

In parallel with the theoretical work of Einstein and Planck, experimental physicists had begun to explore the structure of atoms by the end of the 19th century. In 1895, based on the experimentally well-established facts that atoms as a whole are electrically neutral and that negatively charged particles (electrons) can be knocked out of them, the English physicist Joseph John Thomson developed the first physical atomic model. He imagined the atom as a positively charged sphere in which the electrons are embedded like raisins in a cake. This raisin cake, however, was anything but massive, because physicists already knew that there is a great deal of empty space in the atomic world.

> In Thomson's atomic model, positively and negatively charged particles are evenly distributed throughout the atom.

One of Thomson's students, the New Zealander Ernest Rutherford, showed in a memorable experiment (which is known to every middle school student today) that Thomson was completely wrong with his model. Rutherford bombarded a wafer-thin piece of gold foil with radioactive alpha radiation and placed photographic plates around the gold foil to measure how the radiation was deflected by the atoms of the gold foil. As expected, only the photo plate right behind the gold foil recorded massive radiation—the alpha particles passed through the gold atoms almost unimpeded, with only a few alpha particles getting diverted by the foil. On closer inspection, however, Rutherford noticed something strange. Part of the radiation had been deflected by 90 degrees or more, sometimes even nearly 180 degrees. The distribution of deflection angles provided a direct indication of the mass distribution in the atom and the result was clear: the atom could not be a homogeneous sphere. The mass distribution in the atom had to be much less uniform.

> In Rutherford's model, the atom looks like a tiny version of the Solar System: around a nucleus of positively charged particles, which is very small in relation to the total atom, the electrons circle like planets at a relatively great distance.

But Rutherford's model came with a serious problem: according to classical physics, such atoms could not be stable. The analogy between the atom and the Solar System had its limits. In planets, the centrifugal force of the rotational motion and gravity balance one another. However, negatively charged electrons would have to emit radiation during their circular motion, since each circular orbit involves an acceleration, and accelerated charges emit electromagnetic waves. Physicists speak of synchrotron radiation. The electrons would thus slow down and eventually plummet into the atomic nucleus, and indeed, electromagnetic theory suggests that this should happen within a small fraction of a second.

The experimental physicist Rutherford was at his wits end. It was time for the theorists to take over. One of the most brilliant among them was the Dane Niels Bohr. Today, given the associated financial potential, Bohr might have become a professional football player. He only barely missed the draft for the Danish national team, unlike his brother Harald, who was considered one of the best players of his time, playing with the Danish team and winning the Olympic silver medal in the first Olympic football tournament in 1908. Fortunately for science, in 1903, Niels Bohr went to university to study physics. The first years after his studies in Copenhagen he spent in Cambridge and Manchester, where he began his own research. There the young Bohr stumbled upon Rutherford and his new atomic theory, whereupon a fruitful collaboration began between the visionary experimental physicist and the mathematically brilliant theoretical physicist.

Bohr had learned about Planck and Einstein's theory of quanta and wondered if the quantum concept could not also be applied to the atom. For Planck and Einstein, it was the energy of electromagnetic radiation that could only be emitted in quantum form. Bohr now transferred this idea to the energies of the electrons in Rutherford's atom (more precisely, to their angular momentum). In their motion around the atom, he proposed that electrons could only circle on certain discrete orbits where they retained their energy, but that they could also jump from one of these allowed states to another.

> Bohr's idea of the atom finally provided a way round the criticisms from classical physics: electrons circle around the atomic nucleus on defined orbits and make quantum leaps.

The transition from a state of higher energy to a state of lower energy comes with the emission of a Planck quantum or an Einstein photon. Conversely, an electron can jump from a lower to a higher energy state when it is supplied with an energy packet by an incident photon at the appropriate frequency (energy). An electron that is on its orbit around the atomic nucleus can therefore only absorb or emit energy in certain denominations. It then jumps to the next higher or lower level. For a high enough energy, it can also skip several levels at once. Contrary to what the popular usage of the word suggests, a quantum leap is the *smallest possible* jump within an atom.

> The catch is that Bohr's quantum leaps are no better explained by the laws of classical physics.

The ideas physicists developed to understand atoms provided better and better explanations for the results of their experiments. But they came with a fundamental problem: both the model and the experimental results were incompatible with classical physics. In their search for the connection with well-known laws, physicists gradually got drawn deeper and deeper into a completely new world.

Barcodes of the Elements

With his model from 1913, Bohr had also found the explanation for a conundrum that physicists had not been able to solve for several decades: the emission spectra of the elements. In the 19th century, physicists and chemists had discovered that the light emitted by the various elements when heated up had characteristic colour spectra. For example, sodium emits light at a characteristic wavelength of 589 nm, corresponding to the typical orange–yellow of sodium vapour lamps. Since all known elements can be identified on the basis of their special colour code, spectroscopy has since become indispensable in a multitude of scientific fields, from chemical analysis to astrophysics.

Although at the beginning of the 20th century scientists had already been working with the spectral lines for a hundred years,[4] they did not understand the origin of these lines. Where the spectra came from was completely unclear to physicists and chemists. The only successful attempt to approach the spectral lines mathematically had been undertaken by the Swiss school-

---

[4]They were discovered in 1802 by William Wollaston and, independently of Wollaston, in 1814 by Joseph von Fraunhofer.

teacher Johann Jakob Balmer. In 1885 Balmer set up a formula with which the frequencies of the spectral lines could be calculated for the simplest of all atoms, hydrogen. However, he had come across this formula in a rather heuristic way, by simply extrapolating observed data. He had no explanation as to why his formula correctly described the observed spectrum.

With Bohr's model, the spectral lines could be easily explained: since atoms of different elements allow different energy states for their electrons, each element has its characteristic emission spectrum.

> The bar code-like pattern in the spectrum of each element is a direct expression of the quantum nature of the electron energies in the atom.

At this point physics already counted three hypotheses based on ad hoc assumptions of a quantum nature in the microcosm:

- Planck's quantum hypothesis
- Einstein's photon hypothesis
- Bohr's atomic model

Ad hoc assumptions may be useful for identifying a pattern in observations, but they are unable to provide real explanations—sheer horror for theoretical physicists!

> Planck, Einstein, and Bohr must have felt like magicians who do not know why their trick works.

Bohr and Planck in particular were deeply dissatisfied with their quantum hypotheses, which broke so much with the continuity of classical physics. In the early 1920s, about ten years after Bohr's model, quantum physics fell into a deep crisis. What it was missing was a consistent physical theory that was able to explain the essence and nature of these quantum leaps in the microcosm. But the much-anticipated theory of the quantum finally emerged in the five years between 1923 and 1928, as a result of the collaborative effort of many of the most brilliant scientific minds of the twentieth century. It was to be named *quantum mechanics*.

7

Tertium Datur: Wave and Particle at the Same Time

In the early 1920s new experimental discoveries emerged on an almost monthly basis. Thus, in 1922, by scattering X-rays on electrons at the surface of crystals, the American physicist Arthur Holly Compton succeeded in showing that the collision laws that apply to mechanical particles apply equally to electromagnetic radiation. This was another confirmation of Einstein's photon hypothesis. However, other pieces of the experimental puzzle did not fit into the picture. For example, there was the complex splitting of the spectral lines of atoms exposed to a magnetic field, an effect first observed by the Dutch physicist Pieter Zeeman in 1896.

> Some results from experiments fitted well with the new models, and physicists gained confidence. Other results, however, raised new questions, forcing them to make more and more ad hoc hypotheses.

In the early 1920s, a crowd of brilliant young theoretical physicists began to gather around Niels Bohr in Copenhagen. The Dane became the intellectual father of a generation of scientists who wanted to take on the problems of the new quantum theory in a way that would be as independent as possible of the ideas of their predecessors. Many established physicists spoke in a somewhat derogatory manner of the new "boys' physics" (in German, "Knabenphysik"), noting that most members of the Copenhagen group were only marginally older than 20 when they set out to solve the mysteries of the quantum world.

© Springer Nature Switzerland AG 2018
L. Jaeger, *The Second Quantum Revolution*, https://doi.org/10.1007/978-3-319-98824-5_7

Wave–Particle Duality 2.0

One of these young physicists was Wolfgang Pauli, born in 1900. He tackled the question of why the electrons in the larger atoms with many electrons occupy a whole series of different orbits in Bohr's atomic model. Why didn't all the atom's electrons accumulate on the orbit closest to the nucleus? To explain this, Pauli introduced a strange new rule into the quantum world: a quantum state occupied by an electron in an atom (determined by the three quantum state variables energy, orbital angular momentum, and magnetic moment) cannot be occupied at the same time by another electron. In the new language of quantum theory, this meant that they could not possess identical "quantum numbers". If two electrons never assume the same quantum states as they rotate about the atomic nucleus, they cannot occupy the same trajectory either.

However, in each quantum state in the atom, there were always exactly *two* electrons, and not just *one*, as prescribed by the "Pauli exclusion principle". So Pauli was forced to introduce yet another ad hoc hypothesis. He postulated the existence of another state variable that would distinguish electrons with otherwise identical quantum numbers: the electronic spin. This spin could assume two states: "up" or "down".[1] Before Pauli, Samuel Goudsmit and George Uhlenbeck had already introduced such a spin in an equally ad hoc manner to explain the above-mentioned "anomalous" Zeeman effect (there is also a "normal" Zeeman effect that can be explained classically).

> Wolfgang Pauli introduced another state variable into the new physics: the electron spin. It was later proven to be relevant for other particle classes as well.

But it was the same old story over and over again: Pauli's ad hoc hypotheses raised more questions than it was able to answer. How could a particle that did not have any spatial extension have something like an intrinsic angular momentum?

---

[1]The direction of "up" and "down" is not thereby specified. In principle, spins are possible in any direction, as are any linear combinations of "up" and "down" spins. This can then be, e.g., "up" or "down" in a different direction.

In 1924 light finally appeared at the end of the tunnel. The first step was taken by a young French physicist, who put forward another daring hypothesis in his dissertation: if light can behave as a wave as well as a particle, why should the same not apply to matter, concretely to electrons, asked Louis de Broglie. His idea was a bombshell. Experimental physicists got down to work immediately. From wave optics, many experiments were known that displayed such effects as diffraction and interference (superposition) of waves. And indeed, in analogous experiments they were able to show that, in addition to their particle nature, electron beams could also behave like waves (more about the famous double-slit experiment in the next chapter).

One of these experimental physicists was the son of the discoverer of the electron. What irony! Joseph John Thomson (the one who had come up with the first atomic model in 1895) had received the Nobel Prize for the discovery of the electron as a particle, while his son George Thomson was awarded the same prize for proving its wave properties.

> Wave–particle duality is not limited to light. Electrons also display both particle and wave properties.

Physicists now had to explain not only why light can exhibit certain properties of matter, but also why matter can behave like electromagnetic waves. Under some pressure now, they began to realize that the conceptual and perceptual worlds of classical physics, as they now officially called physics before the discovery of quantum phenomena, were no longer sufficient to interpret the new experimental phenomena. The way towards completely new views and concepts was open.

Wave Mechanics—A Brave New Abstract World

Physicists had failed so far in their attempts to find a mathematical description whereby light could occur both as an electromagnetic wave and as a particle. But with de Broglie's new insight, they could begin to tackle the problem from the other side: how can one describe mathematically the idea that a particle behaves like a wave? A template for such a theory existed already: the well-known Maxwell wave theory. Less than three years after de Broglie had postulated the wave nature of the electron, the Austrian physicist Erwin Schrödinger found the long-awaited equation for a wave theory of matter.

The decisive moment of inspiration came to him during a Christmas holiday in the Swiss ski resort of Arosa, where he had intended to spend a few days relaxing and skiing. He did this in female company (we still don't know who with), while his wife stayed in Zurich with her own lover. In these few week, whether inspired by his extramarital affair or by the mountain air, Schrödinger had a stroke of genius which made him one of the greatest physicists of the 20th century. With the wave equation he discovered, Schrödinger was able to explain the behaviour of the electrons in the atom elegantly and exactly, including Bohr's rules. On his return to Zurich, he received help in the precise formulation of his equation from his mathematical colleague and close friend Hermann Weyl, who (with the knowledge and consent of Schrödinger) was the lover of his wife.

The revolutionary aspect of the Schrödinger wave equation was this: its solutions describe the electrons as waves bound within the atom. Just as the free length of a violin string dictates its frequency, i.e., the tone, the length of the orbit determines the wavelength of the electron.

> Schrödinger's wave equation describes the electron orbits in the atom as standing electron waves of a certain frequency.

What the Bohr model described as discontinuous quantum leaps in Schrödinger's wave mechanics became transitions from one electronic state of self-oscillation to another. Now, at last, a lot of things were beginning to fit together. Because electrons bound in atoms can only assume certain wavelengths or frequencies, their energy can only take on certain (quantized) values, otherwise their waves do not "fit" the length of the closed path.

With his wave equation and its solution, the wave function, Schrödinger had found the basis for a theoretical description of processes in the microcosm. The fact that Bohr's particle model and Schrödinger's wave equation yield the same results gave the physicist a tremendous boost of confidence.

Despite Schrödinger's beautiful mathematical description of the De Broglie waves and the characteristic electron movements in the atom, a fundamental question remained unanswered: what are electrons and photons really? And how can their particle and wave natures be reconciled?

Heisenberg's Uncertainty Principle and the New Quantum Mechanics

Physicists were still trying to visualize the atom. Was it essentially matter or essentially wave? The "religious wars" were far from over yet. Schrödinger wanted to completely replace the particle model with a wave model. Electrons should actually move like waves.

But the work of another brilliant young theorist, this time in Bohr's group (to which Schrödinger did not belong) was finally to remove any hope for an intuitively accessible interpretation of the micro world. In June 1925, about half a year before Schrodinger's erotic and intellectual flight of fancy in the Swiss Alps, the 25 year old Werner Heisenberg was suffering from a terrible bout of hay fever and headed off to Helgoland, a secluded and pollen-free North Sea island. It was here that Heisenberg identified the other decisive feature required to construct a consistent theoretical foundation for quantum theory.

So far, in their thinking about the properties of atomic particles, physicists had naturally transferred the familiar and intuitive views of the classical world with its known state variables to the atomic sphere: like cannonballs, electrons should possess measurable properties such as location, velocity, and momentum. But Heisenberg asked himself whether it might not be possible, and maybe necessary, for a proper description of the atom to dispense with the classical concepts of location and velocity for quantum particles? After all, these parameters had never yet been observed or measured. Heisenberg wanted to rely exclusively on values for frequencies, energies, and intensities known from experimental measurements. With this limited set of variables, he aimed at an unambiguous description of the properties of the atom. He was no longer interested in an intuitive interpretation, but only in computational comprehensibility and describability of the experimental results.

> Werner Heisenberg was the first physicist to break away from the idea of having to "grasp" the atom. He laid his stakes 100% on mathematics.

The young physicist had to establish a connection between observable quantities such as frequency or energy and unobservable variables such as position and momentum. For this purpose, Heisenberg was forced to introduce new rules of calculation that replaced the classical computational links between the various measurable physical variables (for example, simple addition and multiplication) by more complicated computational methods. In the mathematical treatment Heisenberg used, measurable physical variables became functions, so-called operators, which act on abstract states. To his delight, his method allowed him to calculate all the properties of atomic systems, from their stationary energy states to the emission and absorption of photons. On the morning after the night he made his breakthrough, he was so excited that he climbed Helgoland's most distinctive rock, the Lange Anna, on a breakneck ascent. It was lucky that he was a skilled climber, otherwise the Heisenberg uncertainty principle would probably bear a different name today.

When Heisenberg returned from Helgoland full of optimism and euphoria and presented his calculations to his colleagues, Max Born and his collaborator Pascual Jordan immediately realized that Heisenberg's computational methods were actually matrix multiplications, i.e., Heisenberg's operators could be expressed as matrices. This is a computational method from linear algebra, today familiar to every student of mathematics, but which at that time was still relatively unknown. For the first time, the "matrix mechanics" formulated by Heisenberg, Jordan, and Born enabled physicists to bring some order into the confusing disarray of ad hoc hypotheses within quantum theory. With its help, they were also able to mitigate wave–particle duality to some extent.

However, there was a frog to swallow. Heisenberg's matrix mechanics entailed a fundamental new principle of quantum theory: the position and velocity (or equivalently, momentum) of a particle *cannot* be determined to arbitrary accuracy at the same time. Heisenberg had realized that the unobservability of the exact trajectory of an electron does not come from inadequate experimental means, but is a direct consequence of the theory itself. This is the content of Heisenberg's most famous formula, his "uncertainty principle": $\Delta x \cdot \Delta p \geq h/4\pi$, where Δp and Δx denote the inaccuracies in the measurements of the momentum and the position, and h stands for Planck's constant.

Heisenberg stated this as follows: the more precisely we want to determine the position of an electron, the less we can know about the momentum or velocity of the particle, and vice versa.

Wave Function and Probabilities—The Departure from Physical Causality

Experts were now confronted with two new theories of the atomic world: Heisenberg's matrix mechanics and Schrödinger's wave mechanics. Which view was the right one?

A fierce competition emerged between the two physicists, and it even took on a personal note. The camps of their respective followers were deeply divided. Most physicists preferred Schrödinger's interpretation. It was less abstract and more in line with classical ideas. Einstein called Heisenberg's matrix mechanics a magic square (Hexeneinmaleins, literally "witch's one by one"), as it involved a much higher degree of abstraction.

> A stalemate was reached between the views provided by Heisenberg's matrix mechanics and Schrodinger's wave mechanics. In the end, it was the German physicist Max Born got things moving again.

In the same year in which Schrödinger published his equation, Max Born formulated a completely new interpretation of the Schrödinger equation and its solution, the wave function. According to Born, the Schrödinger waves do not describe the physical motions of the electrons per se; rather, the wave function, or more precisely its squared modulus, determines the distribution of *probabilities* for the electron to be at a given location at a given time. In the presence of many electrons, the probability distribution becomes the empirical distribution, where the whole ensemble of electrons is observed as a wave. Considered individually, however, they remain particles.

Here we recognize Einstein's idea about light particles twenty years earlier. In large numbers his light quanta could equally explain the wave nature of light without giving up the individual particle properties. Analogously to electromagnetic waves, which were supposed to correspond to large ensembles of light particles, Schrödinger's waves could be considered to represent space-time averages of a large ensemble of electrons. But while Einstein had still been using the classical methods of statistical mechanics, Schrödinger's equation and Born's interpretation of it represented entirely new physics.

> With Born's interpretation, the Schrödinger equation only allows us to determine the probability of an electron being at a particular place.

According to Born's interpretation, one cannot make any statement about where precisely on its orbit an electron is located at any given time. Can we then even talk about a particular path on which an electron moves? By now it was becoming clear that the Born–Schrödinger theory and Heisenberg's concept were not as far apart as physicists had originally thought. Strictly speaking, Schrödinger had already abandoned the idea of an electron orbit and replaced it with eigen oscillations of electron waves. With these particle waves, the electron could no longer be said to have an exact position, exactly as in Heisenberg's interpretation.

> At second glance, the physical interpretations of the uncertainty principle in Heisenberg's matrix mechanics and Schrödinger's wave mechanics proved to be compatible.

Schrödinger's equation, Heisenberg's matrix mechanics, and Born's interpretation constituted the decisive elements in the interpretation of atomic phenomena. Soon after the publication of matrix mechanics and his own wave mechanics Schrödinger was able prove that the two approaches are also *mathematically* equivalent: Schrödinger's equation can be derived from Heisenberg's theory, and conversely, Heisenberg's matrix mechanics can be deduced from Schrödinger's equation.

There was yet another link between the two views. Heisenberg had derived the uncertainty principle directly from his own matrix mechanics. The uncertainty of position and momentum means that the system can no longer be exactly described in terms of both variables. The catch is now that this principle can also be deduced from Schrödinger's wave theory and its equation. In wave theory, the position–momentum uncertainty corresponds to the well-known fact from optics that a wave cannot be resolved with arbitrary accuracy in terms of both its position and its frequency at the same time. The more accurately the frequency is determined, the less accurately the position can be known and vice versa.

Conversely, for Schrödinger's equation, this meant that with the aid of Heisenberg's uncertainty principle, it could also be interpreted statistically: the uncertainty in position and momentum means that the system can no longer be described precisely deterministically. Although they preferred Schrödinger's approach, physicists now had to work with the "Heisenberg" probabilities in the form of the position–momentum uncertainty.

> Heisenberg's matrix mechanics and Schrödinger's equation had proved to be equivalent. This correspondence made physicists increasingly trusting of the new theory, which they now commonly referred to as *quantum mechanics*.

But the price they had to pay was high. With their new tools, physicists had given up any reference to a concrete intuitive interpretation of the micro world:

- The position and momentum of a particle can only be determined statistically, i.e., by specifying probabilities. This is a fundamental break with a world view that was central to physics since Galileo and Newton, according to which physical systems develop deterministically. This meant that anyone who knows the precise values for position, velocity, mass, air resistance, etc., of a cannonball can calculate exactly where it will go. In quantum physics, on the other hand, the behaviour of even a single electron can no longer be determined exactly, even with the best possible knowledge of all its variables. Although the wave function of quantum system has deterministic dynamics (Schrodinger's equation is a deterministic differential equation), it is not directly measurable.
- One had to accept that an electron could no longer be described in the intuitively accessible three-dimensional space, but only in the abstract state space of the wave function. Mathematically, this state space is a space with infinitely many dimensions. Mathematicians call it a *Hilbert space*.
- Finally, the values of the wave function are not limited to the real numbers we are so familiar with. In fact, it is a function taking values in the set of complex numbers. These have the form $z = x + iy$, where x and y are real numbers but i is the "impossible" square root of -1.

> The laws of quantum mechanics meant turning away from certainties that had been the basis of science for centuries.

Schrödinger himself was highly dissatisfied with this development. Other physicists such as those in Niels Bohr's group in Copenhagen, on the other hand, willingly accepted Born's interpretation and tried to develop the theory further, hoping that all the other pieces of the puzzle could be integrated into it.

Physicists were now happy to be able to describe atoms in a mathematically consistent way. They were also pleased to note that calculations based on quantum mechanics provided results that exactly matched experimental outcomes. But neither Heisenberg's matrix mechanics nor Schrödinger's wave theory could address the general uneasiness when it came to wave–particle duality. How should they do justice to the wave character and at the same time to the particle nature of matter?

With the Heisenberg—Schrödinger theory, the debate about the appropriate interpretation of quantum phenomena and wave–particle duality was not yet over. On the contrary, it had only just begun.

8

As Well as and Neither/Nor: Superposition—How Things Can Be Here and There at the Same Time

With Schrödinger's wave theory, Born's probability interpretation, and Heisenberg's uncertainty principle, physicists were finally able to calculate the phenomena in the microcosm mathematically and interpret them to some extent physically. But physicists were not entirely happy with quantum mechanics. This was in particular due to one of the many peculiarities of quantum particles that completely contradict our everyday intuition: quantum mechanical states can only be assigned probabilities.

This may not sound that dramatic at first. In our everyday lives we often have to be satisfied with probabilities. However, there is a big difference between the probabilities in our world and those of the micro world. For example, if there is a table in one of two locked rooms and we do not know which room it is in, we can only guess where it stands, and we will get the answer right with a probability of 50%. But regardless of whether we open the doors and look or not, it is a reality that the table is in one of the two rooms. Only our *subjective* ignorance about an established fact forces us to deal with probabilities. With a measurement (we open the door and look), we learn where the table stands (and also stood before). Nothing changes within the system itself when we take a look.

In quantum mechanics, on the other hand, probabilities are an *objective* ingredient of quantum mechanical dynamics. This already becomes clear with Heisenberg's uncertainty principle: the limitation on how accurately we can measure the position and momentum of a particle is an *inherent* property of quantum objects. If the table above behaved like a quantum particle, it would not be our lack of knowledge that made us uncertain about where

© Springer Nature Switzerland AG 2018
L. Jaeger, *The Second Quantum Revolution*, https://doi.org/10.1007/978-3-319-98824-5_8

it is. Before we take a look, the table would be so to speak in both rooms at the same time—and equally in neither of them. Only at the very moment in which we take a measurement (open the door and look), would the "quantum table" manifest itself—one could also say materialize—in one of the two rooms.

> The state of a quantum system and its properties are objectively undetermined. It is only with observation that they receive their characteristics.

This means that a quantum object is not at a particular position at a specific point in time. It is everywhere and nowhere. It moves along different paths at the same time, and thus stays in different places at the same time. More generally, it is in a quantum mechanical state in which several classically exclusive properties overlap.

These state properties include not only the position of the quantum object, but also whether it behaves like a wave or a particle, which spin it possesses, and so on. Which of these properties it will manifest upon observation cannot be known before the measurement process. And this not because we just do not know before the measurement what properties the particle has, but because they are *actually* objectively indefinite. Only with the measurement does a quantum particle enter a certain state with clearly defined and distinct properties. Prior to the measurement, it is in an overlap of many different states with their respective properties. Physicists say that the particle is in a *superposition* of many possible states.

> Superposition means that an electron, before being measured, exists as an overlay of different states (for example, places).

Bizarre Behaviour at the Double Slit

Particularly impressive is the quantum mechanical feature of superposition in the so-called double-slit experiment. In this experiment, an electron beam is fired at a screen with two narrow parallel slits. A certain proportion of the electrons passes through the slits and falls on a photographic plate mounted behind the panel. Each impinging electron leaves a black dot on this plate. (Of course, this experiment can also be carried out with other quantum objects, such as photons.)

At first the electrons do not seem to behave differently than macroscopic particles. Because those that make it past the screen have evidently flown through either the left- or the right-hand slit, the blackened areas behind each of the two slits gradually turn into two black stripes. After a sufficient number of such hits, however, there appears a significant difference: the electrons on the photo plate generate a pattern of several alternating stripes of black (electrons landed there) and white (no electrons landed there). Such interference patterns are well known from wave optics and can also be observed with light and water waves passing through a similar set up.

With Born's (or Einstein's) interpretation of the electron probability waves, this pattern can easily be explained, too: on their way through the slit and behind it the electrons behave like waves. It is only when they hit the photographic plate that their particle nature becomes apparent. At that moment, it is decided where exactly on the photo plate a single electron leaves a black spot. Before that moment, its location is fundamentally unknown, and where it ends up hitting the plate cannot be predicted because a wave is not pointlike, but distributed in space. At points where the *wave* properties of the electrons passing through the two slits add up, the *particle* properties of the electrons are more likely to be encountered than where their waves nearly extinguish each other.

In the double-slit experiment, the interference pattern recorded by the photographic plate shows that the electron ensembles behave like waves while passing through the aperture, and like particles only when they hit the plate.

So far so good. But when physicists did the same experiment with single electrons passing through *one after the other*, they were in for a big surprise. Although it was now impossible for several electrons and their waves to interact with each other after passing the two slits, interference effects still occurred. There are only two explanations for this:

- Individual electrons "know" within the collective how to distribute themselves.
- Each electron passes through both slots at the same time and then interferes with itself.

The former seems even crazier than the latter. Physicists thus concluded:

Not just an entire ensemble of electrons, but even a *single* electron can behave as a wave. It passes through both slits simultaneously and behind the screen interferes with itself as a wave would do.

But what happens when along the way the experimenter checks which way the single electron actually chose to go? Following our intuition we would expect a laser beam positioned *behind* one of the two slits, which can detect the individually emitted electrons as they pass, to measure only half an electron at a time. Is that even possible? The physicists who performed exactly this experiment were in for yet another surprise:

- If no observation is made between the double slit and the photo plate, the electron behaves like a wave and interferes with itself as soon as it passes through the double slit. Numerous repetitions therefore result in the well-known interference pattern of several stripes on the photo plate.
- If one measures through which slit the single electron has passed, between the slit and the photo plate, one obtains a clear result: it passed through *either* the left *or* the right slit—not through both slits simultaneously. It is as if merely observing the electron destroys its wave nature and makes its particle nature reappear. This is also supported by the fact that (with a sufficient number of emitted electrons), two clearly delineated stripes appear on the photo plate, and no interference pattern is formed.

But here is an even stranger question: how does the electron know that it will be measured *later* and should therefore behave like a particle at the slit? And how does the electron which is *not* measured later know that it should behave like a wave at the slit? If the measurement only changed the properties of the electron, the measurement *after* the slit should no longer change its properties *at* the slit.

Through the experimenter's decision to observe the path of the particle after it has passed through the slit, the quantum object is forced to pass *retroactively* as a particle through either the left or the right slit.

And now we come to something completely crazy. Once the particle is measured after passing the slit, so that the formation of an interference pattern is precluded, the information received by that measurement can be deleted again by a special experimental setup. The experiment now looks like this:

- The single electrons are measured after passing the slit.
- Through that measurement they lose their wave character. Without further intervention, two separate black lines would appear on the photo plate corresponding to electrons passing through one of the slits.
- Using a special experimental setup the information obtained during the measurement is destroyed again before it can be forwarded to the observer. Physicists speak of a quantum eraser in this context.
- As a result, the electron behaves like a wave again, and this causes an interference pattern to be observed.

Such quantum erasers can be realized for photons, for example, by means of suitable crystals, which determine the path the particles took in the double-slit experiment, followed by a polarization filter (only letting through light with certain oscillation directions), which deletes this information again. Because the results of these experiments are so bizarre, here is a summary:

Case 1: The information is retrieved and the observer will know which way each electron has gone. The electrons behave like particles and two distinct stripes become visible on the photo plate, one behind each slit.

Case 2: If this information is erased before the particle reaches the screen, then once again the observer does not know which slit the electrons have gone through. The result is an interference pattern that corresponds to the wave nature of the electron.

By subsequent destruction of information, the change made to the quantum object by the measurement is reversed again.

It seems as if it is not the measurement itself, but our *knowledge* of whether or not there exists a measurement result that determines the properties of the electron—and this retroactively in time.

The Mysterious Collapse of the Wave Function

Superposition has even more amazing effects in store. There is, for example, the moment when the electron hits the photo plate. If a single electron is registered, it ceases to exist as a wave, and inevitably the probability of the electron striking at any other part of the photo plate instantly becomes zero.

But just before the electron falls onto the photographic plate, the electron wave still exists as a superposition of different positional states. The probability of measuring the electron at any other place on the photographic plate at that time is therefore different from zero.

> The electron retains its superposition, and thus its potential to be measured anywhere in the room, right up until the moment when a position measurement is made on it, or when it hits the screen.

Quantum mechanically, the impact of the electron on the photographic plate means that the wave properties of the electron instantly disappear and only its particle properties remain effective. The wave function of the electron collapses and is now concentrated at a single point. Without any time delay, it suddenly becomes impossible for the electron to hit elsewhere. Physicists speak of an (instantaneous) "collapse of the wave function".

> When the wave function collapses, the superposition disappears instantaneously. The electron, previously distributed as a wave in space, leaves a well defined spot on the photo plate.

Bohr's Principle of Complementarity—The Copenhagen Interpretation

What exactly happens at the double slit? Niels Bohr's answer was that particles and waves are classical terms that cannot be applied to atomic particles. An electron can be regarded as a quantum object and, as such, possesses no equivalent in our normal, classical world of intuition. Therefore, the electron can behave *like a wave* and equally *like a particle*, just not at the same time. Bohr speaks of the "complementarity of particles and waves". To give a full description of physical processes at the atomic level, both views are necessary and complement one another. In honour of Bohr and his colleagues, this interpretation of the quantum physical superposition is called the "Copenhagen interpretation" of quantum mechanics.

> According to the Copenhagen interpretation, an electron is both a wave and a particle, but also neither of the two, and never both at the same time.

The Copenhagen interpretation gives an explanation for the idea that a physicist asking for the nature of the electron must choose for himself what he wants to measure:

- If he interprets (and measures) a quantum object as a particle, he must use the discrete particle model of Heisenberg and resort to its matrix mechanics.
- If he takes it to be a wave, Schrödinger's equation and its statistical interpretation will apply.

In describing and predicting the behaviour of quantum objects, the two theories yield the same results. Only one thing is impossible: to measure both properties, wave and particle, at the same time. Because before the measurement, the properties of the electron are not objectively determined.

> Measurements can never cope at the same time with the wave and particle nature of quantum objects, because in themselves and objectively (independently of the measurement), such objects have neither property.

With their Copenhagen interpretation, Bohr and Heisenberg also gave an intuitive physical explanation for the quantum-physical uncertainty that Heisenberg had identified. If an observer wants to measure an electron, he has to accept that it will interact with his measuring instrument. The electron cannot be measured in isolation, just by itself, because the measuring system must always be included in the quantum system, and the two must be considered as a combined system. This fact is often described by the statement that the observation or measurement process disturbs the system and thus alters its properties, but this statement is misleading, since quantum systems have no independent properties before the measurement.

It's a bit like watching a single snowflake falling from the sky. Once it falls on our palm, it melts due to our body heat. What we see is a tiny drop of water. We do not know what that water drop was 10 s ago: solid (like a snowflake) or liquid (like water)? On the other hand, when we measure a macroscopic system, the interaction with the measuring system does not normally play a significant role. The weight, temperature, density, etc., of a whole snowball can be determined exactly (or at least, to within the accuracy of the measurement device), and the warmth of the palm is thereby negligible.

It is precisely this in principle inevitable inaccuracy of measurements in the quantum world that is the physical manifestation of Heisenberg's uncertainty principle. It provides the mathematical formulation of what physicists have found in experiments:

- the wave and particle property,
- the position and momentum,
- and also the time and energy

of an electron cannot be resolved to any desired degree. The two components of each of these pairs of properties remain blurred.

> The basic idea of Heisenberg's uncertainty principle is that not all the properties of a single electron can be determined to any desired degree at the same time. The attempt to determine one of these variables already changes the other.

However, this limitation is *not* the result of limited or inherently inaccurate measurement tools, but lies in the nature of the quantum object itself. It should be borne in mind that all these properties of the particle are not just indeterminate before the measurement; they *do not exist* before the measurement—just as the "quantum table" mentioned at the beginning of this chapter does not exist in either room unless someone opens the door and takes a look (and neither is it not there).

Disagreement—The Bohr-Einstein Debate About a Spooky Action at a Distance

The Copenhagen interpretation of quantum physics was controversial from the start. Its most prominent opponent was Albert Einstein, who had difficulty accepting the following ideas:

1. quantum mechanics no longer allows an objectively independent existence for particles,
2. as a matter of principle only statistical statements about the atomic world are possible.

Although Einstein recognized the advantages of quantum mechanics with its ability to provide mathematically accurate descriptions and calculate atomic phenomena, he remained convinced throughout his life that the Copenhagen interpretation was incomplete. For him, it was not possible for quantum objects to have no real and independent characteristics, and hence no objectively measurable properties. He therefore felt that there had to be fundamental physical properties that physicists had not yet discovered (so-called hidden variables), and which could codify the true nature of quantum entities in an objective manner. For him, a quantum theory was acceptable only if one could assign exactly one theoretical counterpart in the form of a variable to each element of physical reality.

Einstein adopted the position of physical realism. It was difficult for him to let go of the classical framework of thought in which every physical object possesses objectively definite properties at any given time.

And yet a third point worried Einstein:

3. When the wave function collapses, there appears to exist a long-distance relationship that allows instantaneous transfer of information.

This was how he viewed things. After the electron has gone through the slit, it could theoretically leave a mark anywhere on the photo plate. At the moment the electron hits a certain place, the wave function collapses. So all the other parts of the photo plate must be given the information that the electron can no longer exist there—*one* electron leaves *one* black dot. This information transfer would have to take place timelessly, or in physicists' language "instantaneously", and would thus take place infinitely fast.

However, such an instantaneous action at a distance contradicts Einstein's special theory of relativity, which does not allow any higher velocities in the transfer of matter or information than the speed of light. It is thus no wonder that Einstein did not warm to Bohr's interpretation of quantum events. Einstein did not think it possible that widely separated components of the wave function could be linked by a mechanism that operated without any time delay. He pejoratively referred to on-local effects such as the instantaneous collapse of the wave function as "spooky".

The instantaneous decay of the wave function postulated by the Copenhagen group became a bone of contention among quantum physicists.

These disagreements over the interpretation of quantum theory led to the famous Bohr–Einstein debate at the Solvay Congress in 1927. Bohr responded to Einstein's argument about "spooky action at a distance" by saying that the collapse of the wave function did not constitute an action at distance, but simply concerned the character of the wave function as a probability wave. If a physicist measures where the electron is, he knows at that precise moment that it is nowhere else than where he measured it—and the electron itself also "knows" that.

To refer again to the example with the two tables described at the beginning of this chapter: Bohr said that the moment an observer opens one of the two doors and discovers the table, or discovers that it is not there, he knows what is in the other room. He does not have to look there and no transfer of information between the two rooms is required. The interaction between the measured system and the measurement device is sufficient to trigger the quantum physical collapse of the wave function. In this context, physicists speak of a "state reduction of the quantum system" caused by the measuring process.

Bohr further argued that, according to the uncertainty principle, the precise position of a single electron at a given time cannot be known. In fact, the precise time when an electron strikes the photographic plate is uncertain as a matter of principle and cannot be specified exactly. There can thus be no talk of an instantaneous information transfer under these circumstances.

The two greatest physicists of the twentieth century continued their debate on the interpretation of quantum mechanics to the end of their lives—stubbornly, but always with deep affection, respect, and amicability for each other. The issue at the core of their struggle had a serious impact on the further development of quantum physics, which is why we will follow up on it in the fifth part of the book.

The phenomenon of superposition reveals one of the fundamental properties of the quantum world: non-locality. The fathers of quantum theory failed to agree on the conceptual problems associated with it.

9

Loss of Identity: The New Reality Concept of Quantum Physics and Its Consequences

In the classical sense, reality means that things always have unique charac-teristics, independently of their environment or our perception of them. But according to the Copenhagen interpretation, quantum objects have neither objective properties nor independent existence. What we humans can grasp of them, with our limited imagination, they obtain only through interac-tion with their environment. They do not possess a reality of their own, just a potentiality. The latter only turns into (an a priori unpredictable) reality when a quantum object interacts with a macroscopic object such as a meas-uring device. In the jargon used by philosophers this means that quantum objects possess no substantial form of being of their own.

> For Niels Bohr and the Copenhagen interpretation, reality is no longer a mean-ingful concept. A quantum system that is independent of the observer does not exist.

Bohr and his followers did not want to waste time desperately searching for well defined properties of electrons and other particles. Their efforts were solely aimed at formulating observation and measurement results mathemat-ically and recording correlations. They asked what difference it would make if the smallest particles that make up our world did not exist independently beyond observation and measurement? Many interesting physical conse-quences resulted from the abstract mathematical representation of the new concept of quantum physical reality.

© Springer Nature Switzerland AG 2018
L. Jaeger, *The Second Quantum Revolution*, https://doi.org/10.1007/978-3-319-98824-5_9

Resolving the Borders between Subject and Object

In our everyday lives there is a clear separation between subject and object. When a person measures the water temperature with a thermometer, the roles are clearly defined: the subject observes or measures the object, and these exist independently of each other.

In our everyday world, we are also used to the fact that the components of a more complex whole can be broken down into their individual parts and observed individually. For example, a leaf that hangs from a tree in the forest like a hundred thousand other leaves has certain characteristics (size, weight, colour, age, and so on). If a person looks closely at this leaf, nothing changes in its properties, or indeed those of the person. It can therefore be considered as autonomous and independent of its environment. This means that, even if what is observed is only a small part of a large ensemble, the separation between subject and object is preserved.

> The sharp boundary between the observing subject and the observed object is self-evident to us in the macro world.

It quickly became clear to physicists that this separation is no longer valid in the quantum world. Quantum objects have no independent existence and objective identity. As already stated, only observation by a measuring subject can give them their properties. And nor can they be isolated from neighbouring quantum objects and independently examined as individual components of a larger quantum system.

This resolution of subject and object is also evident in the mathematics of quantum theory: the moment a quantum system is measured, the measuring system becomes an integral part of the observed system. The measuring system and the components of the system to be measured are inseparably merged. Mathematically, all components involved eventually combine into *one single* wave function. In the terminology used by physicists: "The overall state does not separate into its component states."

> In the quantum world, there is no boundary between subject and object. All particles involved in the measuring system and the measured system are intrinsically connected with each other and are no longer distinguishable as individual parts of a whole.

Two Electrons—And yet One and the Same

Using the example of hydrogen and helium atoms, whose shells contain only one or two electrons, respectively, it becomes clear what it means when particles are indistinguishable as individual entities and are components of a *single* wave function.

Strictly speaking, every wave function of a particle belongs to an infinite-dimensional space, the so-called Hilbert space. The three-dimensional space in which we humans find our way has three basis vectors, which we describe as "height", "width", and "depth". They span the room, so to speak. An infinite-dimensional space, on the other hand, requires infinitely many of these basis vectors to span it, and every wave function is a vector in that space.[1] While a point in a three-dimensional space needs three coordinates to defined it, a wave function, strictly speaking, possesses infinitely many coordinates.

A single electron fired at a photo plate, for example, can still be viewed and visualized in three-dimensional space, and this also applies to the state of a solitary electron in the shell of a hydrogen atom. Their state space is infinite-dimensional, but its wave function is a function of the three coordinates used to describe space.

However, things can quickly become more complicated. Even the two electrons of the helium atom can no longer be described as two separate wave functions in ordinary three-dimensional space. Rather, the wave function of the two electrons is a single wave that evolves in a six-dimensional space. Mathematically, the helium atom even requires the combination of two infinite-dimensional spaces. Mathematicians call such a combination of two individual infinite-dimensional spaces (one-particle Hilbert spaces) a tensor product.

This all may sound very abstract, but in principle the calculation of such wave functions is a perfectly standard procedure for physicists; the limiting factor is the computing power of computers. Even in Bohr's day, physicists were not afraid of performing the necessary calculations with wave functions. And they had already been familiar with handling higher-dimensional spaces from classical physics because, in the theoretical description of classical many-particle systems, the relevant space (physicists speak of phase space) is also higher-dimensional.

---

[1]A vector is generally an element of a given space, the associated vector space. Most people are familiar with the two- or three-dimensional vector spaces of our intuition. But completely different vector spaces with higher—or even infinitely many—dimensions can be defined mathematically.

> Although the dynamics of the wave function of a multi-particle quantum system exceeds our intuitive and conceptual powers, it can in principle be calculated quite easily.

What was completely new, however, was that the wave function intrinsically interconnects the various quantum objects—the state of the individual components of the measuring system and the observed quantum system are inseparably linked into an overall state in which the various components lose their individual identity.

Mathematically, this holistic aspect of quantum systems is expressed by the fact that the wave function of the total system $\Psi(\mathbf{x}_1, \mathbf{x}_2, \mathbf{x}_3, \ldots)$ is a function in which the individual state variables $\mathbf{x}_i$ (e.g., position vectors of the particles) no longer stand alone and cannot therefore be considered separately from all others (Ψ, by the way, is pronounced "Psi"). Just as individual particles lose their individuality physically speaking, so they do in the mathematical formulation. This has dramatic consequences: in both "worlds", the experimental world and on paper, the components of the entire system can be interchanged without changing any measurable aspect of its overall state.

The identical particles involved in the overall system—in our example the two electrons of a helium atom—thus have no identity of their own and are absolutely interchangeable. They cannot be distinguished from each other in any theoretically imagined way by any of their properties—as quantum objects they have none. (Particles of different types, for example the electron and the proton in the hydrogen atom, are of course distinguishable, and interchanging them does indeed have physical consequences.)

> As constituents of a single wave function, the two electrons of the helium atom are not independent particles. Instead, they are part of a single whole, without any properties of their own, and are thus fundamentally indistinguishable.

Two and yet One

The indistinguishability of identical particles, as harmless as it may at first appear, has some dramatic consequences: for it contradicts a classical philosophical principle formulated by Gottfried Wilhelm Leibniz (which was, however, already known to the Greeks in antiquity): the *Principium identitatis indiscernibilium* (Latin for the "principle of identity of indiscernibles", often abbreviated as *pii*). It states that there can never be two objects that are in every way completely identical. In other words, there are no two apples, people, tables, etc., in the world that are indistinguishable. Conversely, we can only distinguish individual objects in our world from one another because this principle applies. According to Kant, the locality of objects plays a major role here. Even if two things coincide most profoundly in their innermost qualities, they are not identical, since they are to be found in different places.

And that is precisely why the *pii* has no meaning in the micro world. The two electrons of a helium atom share the same mass, charge, and all other physical properties—as well as the spatial property of location, because they share with each other the space in which they find themselves with a certain probability. They even share the spin. Pauli had introduced spin to distinguish the two electrons of the helium atom, but this was merely a somewhat inaccurate attempt to get a hold on states in the quantum world. In fact, the electrons in the helium atom can no more be individually assigned a definite spin than 100 euros in a bank can be associated with a specific banknote.

In the quantum world, the spin wave function of the two helium electrons is a superposition of both combinations: "particle 1 with spin up, particle 2 with spin down" and vice versa, "particle 1 with spin down, particle 2 with spin up". The statement that electron A has spin "up" and electron B has spin "down" is just as true as the statement that electron A has spin "down" and electron B has spin "up". What is certain is that both spin states are represented.

> The two electrons of a helium atom are absolutely indistinguishable, but at the same time they assume non-identical states, because they each have a different spin at any time, although it cannot be fixed and assigned to either of them.

The notion that two quantum particles can be indistinguishable is not a theoretical gimmick, but has real measurable implications, e.g., for the way matter is composed. Molecular compounds share certain electrons that are exchangeable within the total wave function without any consequences. This plays a major role in the stability (and indeed, possibility) of chemical compounds.

The fundamental indistinguishability of quantum particles also affects the way particle ensembles behave statistically, in a way that runs counter to common sense. In classical physics, each component within an overall system has its own identity, just like every individual ball in the ball bath at

Ikea. You can number them. If you swap component 2198 for component 1935, the new state is no longer identical to the old state.

In the quantum world we need to count differently. Whether electron A is at position P and electron B is at position Q cannot be distinguished from electron A being at position Q and electron B being at position P. Both represent one and the same quantum state Ψ.

> With quantum particles, we have to count in a very different way than we would with objects like apples, people, and tables that we encounter on an everyday basis.

The fact that individual particles within a many-particle ensemble are indistinguishable on principle leads to a fundamentally different statistics in quantum theory than in classical many-particle theory. The following thought experiment will illustrate this.

Two balls with the same momentum collide from opposite directions 500 times. The relative positions of the balls during the collision vary slightly in a random manner, so that they fly apart in different directions after the collision. Now count how many times a ball is deflected by exactly 90°. In this case, according to the law of momentum conservation, the other ball will always fly in exactly the opposite direction. Let us assume that Ball 1 (green) flies to the left and Ball 2 (red) to the right. Because the distribution of relative positions is symmetrical, the red ball will on average fly 90° to the left as often as it flies 90° to the right. Suppose the green ball bounces exactly 20 times 90° to the left. Then, due to the symmetry of the distribution of flight directions, we can conclude that the red ball also flies 20 times 90° to the left (and thus the green ball 90° to the right), giving us a total of 40 times an exact 90° deflection in the collision of two balls. A prerequisite for this result is that we can distinguish the two balls (in this example, by their colour).

If the "balls" are indistinguishable quantum particles, things suddenly look very different. In the case of a 90° deflection, there are no longer two options. Mathematically, what appear to be two possibilities (particle 1 flies to the right, particle 2 to the left, and vice versa) are actually only a single possibility. Thus, with the quantum mechanical counting method, the probability of a deflection of exactly 90° upon a random collision is only half what it is for the case of distinct particles. The human mind automatically protests against this conclusion: "That cannot be! Even if all the balls are painted grey, we will be able to observe a 90° event 40 times." It is exactly this that is so hard for us to imagine: the balls are not just all grey, they are entirely indistinguishable! Instead of imagination, what we need is mathematics in this case. And that in turn is what reveals itself when measurements are made, whether the results fit our world view or not.

In fact, in experiments involving quantum particle impacts, the quantum mechanical counting method has been found to be accurate. However, figuratively, a 90° event was not measured 20 times (half of the 40 events expected in the macro world). The quantum mechanical counting method is even more complicated because there are actually two different types of particles—fermions and bosons—which behave differently, as will be explained in the following. Depending on the particles used, physicists count either 0 (in the case of fermions) or 80 (in the case of bosons) 90° events in the corresponding experiments.

> If in our everyday lives two distinguishable components have two statistical possibilities available to them, these can coincide into a single possibility in the quantum world where particles are indistinguishable.

Old Puzzle, New Solution—The Gibbs Paradox

The discovery of the indistinguishability of quantum particles led to a new quantum statistics. With this it was possible to solve a long-standing problem in classical physics, the so-called Gibbs paradox. When, for example, pure oxygen and pure nitrogen are combined in a single vessel, the thermodynamic measure of entropy in the overall system increases compared to the sum of the entropies of the subsystems before mixing. This is because entropy is a measure of disorder and it is virtually impossible for the two gases to separate back into their previously distinct states. When two identical gases of the same volume are mixed, for example, chemically pure nitrogen from two different chambers (at the same pressure and temperature), physicists expected the entropy in the new combined state to be higher than the sum of the entropies in the separate chambers. For at least in our usual way of thinking, the individual gas molecules can be provided with labels saying which of the two chambers each one originates from. Thus the entropy would be calculated as with "red balls" and "green balls", and as the disorder in the gas mixture increases, its entropy must be greater than that of the separated gases. But all physical measurements showed a different picture: the entropy does not increase in this way for identical gases!

Experiments and the (classical) principles of thermodynamics based on the distinguishability of particles contradict each other.

For a long time the physicists were puzzled about this. Their mistake was to consider individual particles in the gases as classically distinct, and this meant a significant increase in the number of possible (micro) states for any given macro state—which is exactly what corresponds to an increase in entropy. Only if we use the quantum statistics of indistinguishable particles do we obtain the observed result.

Alternation Games with Fermions and Bosons

The indistinguishability of particles and the different statistics in the quantum world have even more, equally real, measurable effects. The physical properties of a many-particle system of indistinguishable particles are completely unchanged upon the permutation (interchange) of two of its particles. Physically, however, only the square Ψ^2 of the wave function of the system is measured (and thus relevant). This means that theoretically there exist two solutions for the behaviour of the quantum mechanical wave function upon permutation: either the entire wave function Ψ remains unchanged or it changes its sign from $+\Psi$ to $-\Psi$. Even if the sign change for the square is meaningless, the question of what the negative wave function $-\Psi$ would look like is not just purely academic.

In fact, nature has realized both types of particles which, in an ensemble, behave differently by a sign under permutations:

1. Particles in which the wave function remains unchanged (symmetric) upon permutation are called bosons. They were named after the Indian physicist Satyendranath Bose, who formulated a first theory of bosons with Albert Einstein. The bosons include photons.
2. Particles in which the sign of the total wave function changes when they are interchanged. Physicists speak in this case of antisymmetry. These particles are called fermions, named after the Italian physicist Enrico Fermi, who developed the first theory of their statistical behaviour. The fermions include electrons.

All particles in nature belong to one of two groups: fermions and bosons.

The anti-symmetry of fermion wave functions contains the deeper reason for Pauli's exclusion principle, according to which two electrons can only assume non-identical quantum states. For if as sign-changing fermions they could be in an identical quantum state, the ensemble wave function would change under permutation from $+\Psi$ to $-\Psi$. At the same time, after the exchange, it would still have to be the same overall state (as both particles are in exactly the same quantum state, and they are indistinguishable). In other words, the following relationship would hold: $+\Psi = -\Psi$. This is possible only for the value zero, which would mean that neither wave function nor particles would be present. So there cannot be an ensemble of entirely identical fermions (in this case, electrons) that are in entirely equal quantum states.

For bosons, there is no change of sign, and $+\Psi$ remains $+\Psi$. The wave function can therefore take any value. The Pauli principle does not apply here, and arbitrarily many bosons can be packed into the same state.

Once again, neither the permutation of fermions nor the permutation of bosons leads to a measurable change in the system. But within an ensemble they behave fundamentally differently. If fermions were apples and bosons pears, kept in many separate baskets, then apples could be exchanged for apples and pears for pears (inside the basket and also from basket to basket) without measurably changing the states of the respective baskets. But only one apple can be put in a given basket, because it will be impossible to add a second one to it. On the other hand, countless pears can be put into one and the same basket without restriction. This fundamental difference between fermions and bosons has been confirmed in many experiments.

> Bosons and fermions behave fundamentally differently in ensembles. Bosons can occupy the same state in large numbers, while for fermions, not even two of them can be in the same state.

The permutation symmetry of particles is somewhat mysteriously related to a feature of quantum particles that Pauli introduced to explain the anomalous Zeeman effect, namely, the spin. For it turns out that fermions always carry half-integer spin (for the electron this is $^1/_2$, while other particles carry the values $^3/_2$, $^5/_2$, etc.), and bosons always come with integer spin (photons take the value 1).

The theoretical justification for this relationship between particle statistics and spin did not emerge until a few years later in a fully formulated quantum field theory. With the so-called spin–statistics theorem, physicists stumbled upon a deeper connection with Einstein's theories, because a proof of this theorem is only possible with reference to his special theory of relativity (see the next chapter).

> Spin is a purely quantum property that can be represented mathematically and calculated precisely, but does not allow for an intuitive interpretation. Fermions always have half-integer spin and bosons integer spin.

Superconductivity, Superfluidity, and the Laser—Useful Macroscopic Quantum Effects

The abstract mathematics of the many-body wave function and quantum statistics lead to many macroscopic quantum effects that can be exploited in important technological applications. First there is the effect of superconductivity. In 1911, the Dutch physicist Heike Kamerlingh Onnes discovered that, below a certain temperature specific to the given material, the electrical resistance of many metals falls below zero. The metal then conducts electrical current without any losses. If this were possible at room temperature, all problems of (electrical) energy transfer would be solved. According to classical physics, however, the phenomenon of superconductivity is actually impossible, since at any (non-zero) temperature, there is always an interaction between the conducting electrons and the atoms in the metal wire (scattering). It is the resulting energy losses of the electrons that cause electrical resistance.

Quantum physics explains what happens in superconducting metals. The arrangement of the metal atoms in the conductor can be compared with a lattice structure. When a voltage is applied, electrons move freely through this lattice. While the self-motion of the atoms in this lattice becomes ever smaller at very low temperatures, due to a certain interaction of the electrons with the positively charged lattice atoms, two electrons, which otherwise repel each other, can join together to form so-called Cooper pairs. While each individual electron has a spin of $+\frac{1}{2}$ or $-\frac{1}{2}$ and thus belongs to the class of fermions, these pairs of two electrons now come with a total spin of 0 and are therefore bosons.

If the temperature of the conducting metal drops below a certain value, two electrons (fermions) form a Cooper pair, with integer spin, which thus behaves like a boson.

Pauli's exclusion principle does not apply to bosons, only to fermions. Thus, a vast number of Cooper pairs can be part of a single common (macroscopic) quantum state. When a voltage is applied, all Cooper pairs move through the lattice as one *homogeneous* body. Of course, a single Cooper pair could be scattered on the metal lattice, but since they all are connected into one shared wave function, all other Cooper pairs would have to be scattered at the same time—but the local forces of the metal lattice would not be sufficient for this to happen. Thus, the energy transfer of a single Cooper pair to the crystal lattice is suppressed and the current of Cooper pairs flows without resistance.

In the state of superconductivity, electrons pair up to form bosons and these pairs then constitute a common aggregate quantum state that moves undisturbed through the crystal lattice of the conductor.

A similar effect is superfluidity. This describes a liquid in which any internal friction disappears below a certain critical temperature. In this state the liquid can even flow through very narrow capillaries without any resistance. Superfluidity is another macroscopic quantum effect in which a macroscopic number of bosons (in this case, atoms of certain helium or lithium isotopes) occupy the same quantum state.

However, the best known and most widely used macroscopic quantum effect is laser light. Here all photons of a light beam oscillate in synchronization (with the same phase), propagate in the same direction, and have the same frequency and polarization. As bosons they can and do assume the same quantum state, just like the electron pairs and helium atoms in superconductivity or superfluidity.

The laser is another macroscopic quantum effect that can only occur due to the indistinguishability of its participating quantum particles and their loss of individual identity.

Open Problems in the Copenhagen Interpretation

Let us return to the Copenhagen interpretation. On closer inspection, it turns out that it comes with another fundamental problem. On the one hand, it eliminated the separation between the observing subject and the observed object. On the other hand, Bohr and his colleagues attributed a priori a *macroscopic* character to the measurement system. Its "superior authority" over the *microscopic* quantum system is supposed to be the reason why the measuring system takes on such an existential and essence-constituting function for the quantum object.

Philosophically trained physicists immediately recognize the flaw in this argument. The Copenhagen interpretation constitutes a two-world theory: here the quantum systems that obey quantum physical laws, there the macroscopic measuring systems that obey the laws of classical physics—without any recognizable connection between these worlds. At the same time, however, the macroscopic measuring system must determine the properties of the quantum system. How does that work? And are classical macro systems not at least composed of atoms, which in turn obey the quantum laws?

Bohr could offer only a very pragmatic answer to this question. In 1920 he formulated the "correspondence principle": for sufficiently large systems the laws of the quantum world—somehow—imply the laws of classical physics. Therefore, for everyday objects, a macroscopic wave function with all its bizarre properties and its probability interpretation deviating so strongly from our everyday experience should not even arise. However, the nature of this transition between the two worlds remained entirely open within the Copenhagen interpretation.

Werner Heisenberg spoke of a sharp separation between the macro and micro worlds:

> In a mathematical treatment of the process, a dividing line must be drawn between, on the one hand, the apparatus we use as an aid in putting the question and in a way treat as part of ourselves, and, on the other hand, the physical systems we wish to investigate.

Bohr and Heisenberg arbitrarily fixed an ad hoc boundary between the world we know, governed by classical laws, and the quantum world, in which the laws of quantum mechanics apply. This boundary was called the *Heisenberg cut.*

Without this Heisenberg cut separating the macro and the micro worlds, quantum theory would ultimately have to be applied to the entire cosmos in the form of a single wave function for the whole universe. But who could then be the external observer who, through their act of observation, "brings forth" the existence of the universe?

The explanatory difficulties of quantum physics were a long way from being fully resolved in the 1920s and 1930s. But most physicists turned away from the problems of the Copenhagen interpretation, because the new theory was so successful for calculating and interpreting the experimental results. But just as garbage deposited in the basement of a tower block eventually begins to smell if it is not taken out, all these problems eventually caught up with physicists. The main problems were two details that the reader has already encountered:

1. Superposition, i.e., the coexistence of properties of quantum objects that are mutually exclusive in the classical world. Among other things, it is responsible for the measurement problem, and ultimately for the fact that subjects and objects can never be completely separate in the quantum world.
2. The impossibility of considering two interacting quantum systems separately. Later, this effect was conceptualized as *entanglement*.

> The vagueness of terms such as "complementarity" (of particle and wave) or "correspondence" (between the laws of the macro and quantum worlds) left several important questions unanswered in the Copenhagen interpretation.

In the fifth part of the book, we will return to these two problems and the way they are treated in quantum physics today. The fourth part, on the other hand, will deal in greater detail with the philosophical, spiritual, and religious aspects of quantum physics. But first we want to continue our journey into the bizarre world of the quantum by making the leap from quantum mechanics to quantum field theory. For there is still much to discover.

Part III

From Quantum Field Theories to a "Theory of Everything"—All Material Dissolves

10

Negative Energies and the Electron Spin: Combining the Theory of Relativity to Produce a New Quantum Theory

The Copenhagen interpretation and the Schrödinger equation left many questions unanswered. There was the problem of interpreting the nature of quantum particles (wave or particle?), the conflict between Einstein and Bohr about the collapse of the wave function during the measurement process, and so much more. The electron spin introduced by Pauli also caused a few headaches:

- Even though the Schrödinger wave equation did not involve spin, physicists knew about its existence. But they had no theoretical explanation for it. The origin and deeper reasons for the electron spin remained open. Furthermore, they were unable to explain the mysterious connection between spin and quantum statistics: why are fermions, with half integer spins, and bosons, with integer spins, associated with different statistics?
- Neither Schrodinger's theory nor the Copenhagen interpretation was able to explain the so-called gyromagnetic factor (or g-factor). This is the ratio of the magnetic moment of the electron to its angular momentum. According to the Schrödinger equation it would have to assume the value 1. However, if Pauli's electron spin was taken into account, the g-factor could no longer be derived from theoretical or mathematical considerations. In the meantime, experimental measurements using the Stern–Gerlach setup found a value of almost exactly two. Where did this value come from?

© Springer Nature Switzerland AG 2018
L. Jaeger, *The Second Quantum Revolution*, https://doi.org/10.1007/978-3-319-98824-5_10

All these unanswered questions were compounded by an even more fundamental *theoretical* problem: Schrödinger's, Bohr's, and Heisenberg's quantum mechanics seemed to contradict Einstein's special theory of relativity.

Einstein's Second Stroke of Genius—The Special Theory of Relativity

Besides his work on the photoelectric effect, in which he introduced the idea of light consisting of particles called photons, Albert Einstein published three more seminal works in 1905. One of them bore the title *On the electrodynamics of moving bodies.* In it Einstein rejected the idea of physical space and time as introduced by Isaac Newton, which corresponded so well with our everyday experience and intuitions. The catalyst for Einstein's considerations was Maxwell's theory of the electromagnetic field, developed fifty years earlier. This states that the speed of propagating electromagnetic waves, e.g., visible light, is finite, and always takes the value of about 300,000 km/s in every frame of reference.

In Newton's classical physics, the velocity of an object relative to a new frame of reference is found by simply *adding* its velocity relative to the old frame to the relative velocity of the two frames. Almost 400 years ago, Galileo Galilei had realized that a cannonball dropped down from the mast of a ship will not fall into the sea, as one might assume given that the ship keeps moving forward as it falls. In fact, it will fall directly onto the ship's deck next to the mast (provided the ship is not accelerating). The reason is that the speed of the ball is added (vectorially) to the speed of the ship. Likewise, consider a train traveling at a speed of 100 km/h. A stone is thrown out of the window in the direction of travel at a speed of 20 km/h. From the perspective of an observer in the station, the speed of the stone will be 120 km/h.

Generally, in order to describe an event from the perspective of a moving system, at each time t, each point at position x is assigned a different point in space x' simply by adding the velocity vector times the time elapsed. In the language of physics, we perform a transformation of the position according to the rules of the so-called Galilean transformation to relate positions in two frames of reference.

But when it comes to the speed of light, things are different. Consider a situation analogous to the train above: a spaceship moves past the Earth at a constant speed of 100,000 km/s. Now the spaceship sends out a ray of light that is observed from the earth. According to the classical calculation, when observed from a position on Earth, the light should travel either at 400,000 km/s when the light is emitted in the direction of motion of the spaceship, or at 200,000 km/s when the light is emitted in the opposite direction. But that is not what happens. The fact that the speed of light always remains constant at 300,000 km/s, regardless of whether the observer's system is moving relative to the source of light or not, was proven experimentally in 1887 (to do this, the speed of light was measured relative to the Earth's motion around the Sun).

> If the classical addition of velocities is no longer valid, we are forced to revise the very idea of velocity.

But what is velocity after all? In everyday life it is a quotient of spatial and temporal separations, for example "100 km/h". Space and time are thus considered as absolute entities, clearly separated into two spheres. Einstein's radical as well as ingenious conclusion was that the idea of a time independent of space and a space independent of time is in fact untenable. The invariance of the speed of light under relative motion of the observer can only be explained if the space and time components are directly connected, whence time becomes the "fourth dimension". This is exactly what Einstein's special theory of relativity achieves.

Mathematically, this means that when we change from one moving system to another which is in motion relative to it, we must transform not only the spatial dimensions as in Newtonian physics, but also the time coordinates. The corresponding transformation rules are somewhat more complicated than those used to calculate the point of impact of Galileo's cannonball or the velocity of a stone as seen from the train station. These rules were first articulated by the Dutch mathematician Hendrik Lorentz and are today called Lorentz transformations.

One of the better known consequences of these transformations is that, in moving systems, time passes more slowly and the lengths of objects are shorter than in non-moving systems. In our everyday world, however, the fastest jetliner moves much too slowly in comparison to light for any of these effects to become significant. It is only at much higher speeds that the connecting door between space and time opens up sufficiently for us to notice them. If we could move almost as fast as light relative to another observer, we would experience a world with completely different spatiotemporal characteristics compared to that observer.

According to Einstein's theory of relativity, space and time are no longer independent entities, but are intrinsically interwoven into a global space-time structure.

Einstein's special theory of relativity predicted yet another surprising effect. When the velocity of a particle increases, so does its mass! This results in a general relationship between the energy of a body and its mass. Einstein writes: "When a body releases the energy L in the form of radiation, its mass decreases by L/V^2" (here V stands for the speed of light, today described by the letter c). This ultimately led to Einstein's most famous formula $E = mc^2$.

Let us now return to the problem of quantum mechanics. All experiments corroborated Einstein's new theory. Would it not then be natural to assume that the space-time structure of the special theory of relativity should also apply in the quantum world? If so, the problem was that the fundamental equation of quantum mechanics, the Schrödinger equation, was still based on the old relationship between space and time, where space is viewed as the "container" of the physical world, and time runs independently of it, as an internal (non-spatial) parameter of motion. Thus means that the Schrödinger equation does not obey the rules of the Lorentz transformation when viewed in different frames of reference. Physicists say: "The Schrödinger equation is not invariant under Lorentz transformations, i.e., it changes its form when position and time are transformed according to the Lorentz rules." In short, "it is not Lorentz invariant."

The problem with the Schrödinger equation is that it is a "non-relativistic" equation, i.e., it is based on the space-time conception of non-relativistic physics.

A New Equation and the Solution of the Spin Puzzle

Although up to then the Schrödinger equation had described the experimental results very well (except for the existence of the spin and the g-factor), a new equation for the electron which was compatible with the principles of special relativity had to be developed. In 1927, in a single stroke of genius, a young English scientist succeeded in establishing just such an equation on the basis of purely theoretical considerations. His name was Paul Dirac. The simple title of his paper was *The Quantum Theory of the Electron*.

> From the non-relativistic (i.e., non-Lorentz-invariant) Schrödinger equation emerged the relativistic Lorentz-invariant Dirac equation.

Dirac's new equation turned things around. Like Schrödinger's, it also enabled the calculation of all atomic properties, including the emission spectra. But there was a lot more to it. Dirac's equation allowed him to deduce the spin without the need for any additional assumptions. The spin hypothesis which physicists had been forced to introduce ad hoc to understand experimental measurements now proved to be a direct consequence of a relativistically extended quantum theory! This is the scientists' dream, when many pieces of a puzzle suddenly combine to form an overall picture. Furthermore, the experimental value of the g-factor followed directly from Dirac's equation. Dirac wrote in 1928:

> The incompleteness of the previous theories [lies] in their disagreement with relativity, or alternatively, with the general transformation theory of quantum mechanics.[1]

And Dirac's equation gave physicists yet another gift. The mystifying relationship between spin and quantum statistics could finally be derived. The proof was provided in 1939 by Wolfgang Pauli and the Swiss physicist Markus Fierz. Central to it is the Lorentz invariance of the Dirac equation. The reason that this breakthrough only came twelve years after the invention of Dirac's equation was that the complex reasoning of a relativistic *quantum field theory* (see next chapter) had first to be worked out. With its help, the spin–statistics theorem emerges directly from Dirac's equation: it says that quantum objects are either fermions with half-integer spin or bosons with integer spin. It is a cornerstone of today's elementary particle physics. All experimental findings over the last 90 years are in complete harmony with it.

---

[1]From Dirac's paper *The Quantum Theory of the Electron*, Proceedings of the Royal Society (1 February 1928).

Within the framework of a relativistic quantum (field) theory, the existence of the electron spin and an explanation of quantum statistics could be derived directly from Dirac's equation.

From Consternation to One of the Great Moments of Theoretical Physics

While the theoretically consistent derivation of the electron spin and the experimentally measured g-factor caused physicists to cheer, another consequence of Dirac's equation made their hair stand on end: it allowed for the existence of electrons with negative energies! The Dirac equation has four spatial wave functions as solutions (so called spinors); when spin was taken into consideration in a nonrelativistic way, there are solutions in the form of a two-component wave function (while the spinless Schrödinger equation had a one-component wave function).

Two components of Dirac's four component wave function describe particles in two different spin states, each with positive energy, while the other two components also describe particles in two different spin states, but with negative energy. Thus, for every quantum state with positive energy +E, there exists a corresponding state with negative energy −E—an impossible, even unthinkable situation in classical physics! That would be like a car moving at negative (absolute) speed—not going backwards, but somehow less than not moving. How should physicists interpret these negative energy solutions of the Dirac equation? Were they supposed to just ignore them?

The fact that according to the Dirac equation there should exist particles with negative energy seemed entirely nonsensical.

Even without going into the mathematical details of the Dirac equation, one can understand where the origin of the problem of negative energies lies, by taking a closer look at Einstein's equation $E = mc^2$. This simple relationship between mass and energy is only valid as long as the particle is at rest. As soon as the particle moves, more terms need to be added to this equation. In general, the following equation holds for the relationship between energy and momentum: $E^2 = p^2c^2 + m^2c^4$. Here, the energy E appears squared, i.e., in the form of E^2. Since the square of a negative number is also positive, such an equation, solved for E, always has two solutions, one with positive energy and one with negative energy. When we change the sign of the energy, *all* signs are reversed, so an electron with negative energy would carry a positive rather than a negative charge.

> When the special theory of relativity is applied to classical physics, we can simply stipulate that only positive solutions are relevant. However, this is no longer possible for the equations of quantum theory. Here both solutions are equally meaningful.

Holes in the Sea

Dirac himself proposed an interpretation of the negative energy states. If there are infinitely many such energy states below zero, then all electrons would gradually have to fall to ever lower energy levels (each with the release of photons). As they do not seem to do that, Dirac assumed these lower energy states were already occupied. Just as in the atom, an electron in the second shell cannot fall down to the first shell, because the available places are already occupied by two electrons.

Dirac assumed that what we consider to be empty space (vacuum) is actually a system in which all states of negative energy are filled up (the so called Dirac sea). This would explain why a single electron in this vacuum must remain in its positive energy state. It is the Pauli principle that prevents the electron from falling below the zero energy line, even if there was some way it could release the energy.

However, the opposite is possible: an electron near the "zero line" with negative energy can pick up a photon and switch from a negative to a positive energy state, leaving a hole in the sea, like a bubble in water. This hole is the positively charged counterpart to the electron. It is then promptly replaced by any positive energy electron that can readily release its energy, fall into the resulting hole, and thus replenish the Dirac sea. The two particles, the "normal" electron and its counterpart, effectively annihilate each other. In total, two photons are released, both of which have an energy equivalent to the mass of the electron (511 keV, according to Einstein's formula $E = mc^2$).

> Dirac interpreted the vacuum as an infinite and completely filled sea of particles of negative energy.

Three years later, Dirac interpreted the particles with negative energy as "anti-electrons", that is, with the same mass and spin as electrons, but with the opposite charge and magnetic moment, and positive energy. What the physicists did not know at that time was that such a particle had already been discovered as early as in 1929. The Soviet physicist Dimitri Skobeltsyn, in his attempts to make cloud chamber observations of gamma radiation as part of the cosmic radiation, had observed particles that looked like electrons, but whose orbits in an applied magnetic field were curved in the opposite direction. This meant that they had the opposite charge to the electron, just like the anti-electron predicted by Dirac. Skobeltsyn, however, did not pursue this phenomenon further—unfortunately for him, as he would no doubt have received the Nobel Prize for his observation. Frédéric and Irène Joliot-Curie also observed early evidence for "positive electrons," but mistakenly interpreted them as protons.

On August 2, 1932, the American physicist Carl David Anderson—with a very similar device to the one Skobeltsyn had used three years before—finally discovered the anti-electron. He called it the *positron*. The experimental proof for the existence of the positron was Dirac's greatest triumph. At the same time, a new branch of physics was born: particle physics.

The positron was the first particle whose existence was predicted theoretically before it was observed. The experimental detection of antiparticles was one of the great moments of theoretical physics.

Patrick Blackett and Giuseppe Occhialini discovered the positron almost at the same time, but they wanted to consolidate their results with further measurements and therefore published their results shortly after Anderson. Together with Schrödinger, Dirac received the Nobel Prize in Physics in 1933, and Anderson in 1936, while Blackett and Occhialini came out empty handed.

The Path Towards the First Quantum Field Theory

With the synthesis of quantum theory and the special theory of relativity, quantum mechanics had outgrown itself. The Dirac equation became the foundation of a quantum theory of the electromagnetic field, "quantum electrodynamics," or as it was soon called, the first "relativistic quantum field theory." With its help, the electron could finally be correctly coupled to the electromagnetic field.

But as always in science, any increase in knowledge leads to new gaps in our understanding. The existence of antiparticles and the fact that energy has no lower limit, in combination with the Heisenberg uncertainty principle, directly led to a fundamental problem: there is no absolute zero line for the energy. Because quantum theory, according to Heisenberg's uncertainty principle, prohibits the precise determination of the amount of energy in a system at a given time, the zero field energy must also be blurred. This means that the field energy at any position and time must continually and spontaneously deviate from zero for short periods of time.

These fluctuations mean that a particle–antiparticle pair can spontaneously come into existence by a particle jumping out of the Dirac sea, and this in turn will give rise to an antiparticle (a hole in the sea). Thus, an electron–positron pair can emerge literally out of nothing, that is, without energy input from the outside. However, the two particles will annihilate each other again within a very short time that is specified by the uncertainty principle, i.e., the resulting electron will jump back very quickly to fill the hole in the Dirac sea once more.

> According to quantum theory, for a very short time, spontaneous particle–antiparticle pairs are formed out of the vacuum. Because they cannot be directly observed (measured), such quantum fluctuations are called *virtual particle pairs*.

Although these short-lived quantum fluctuations do not generate permanent particles, they do influence physical measurements. Virtual particles can be measured, for example, in the "Casimir effect". Here they cause a (very weak) force to act between two conductive plates placed in parallel in a vacuum. In theory, this effect had been known for a long time, but it was only in 1998 that it was actually measured in an experiment.

It should be mentioned at this point that the Dirac sea is merely an attempt to illustrate something that is actually more or less impossible to illustrate. Today, the concept of the Dirac sea is long outdated. But the virtual quantum fluctuations do actually exist. In fact, they are omnipresent. To describe them theoretically, a more fundamental theoretical framework than quantum mechanics is needed: quantum field theory. This will be the subject of the next chapter.

11

Quantum Field Theories: All Matter Dissolves

In 1930, there were two quantum theories for elementary particles like electrons: the non-relativistic one due to Schrödinger and Heisenberg and the relativistic one devised by Dirac. Both describe how particles can also behave as waves. On the other hand, there was no quantum theory that had its origin in waves or fields; the electromagnetic field theory had remained classical until then. This was an inconsistency that was hardly to the taste of the theoretical physicist. In addition, the prevailing quantum theory still used the classical ideas of particles described mathematically as points and waves described mathematically as fields. The issue of wave–particle duality—actually, field–point duality—was never entirely resolved in quantum mechanics.

After the spectacular experimental discovery of Dirac's antiparticles, the theorists wanted to transform classical electromagnetic field theory into a quantum theory. In order to fully understand the wave–particle duality, the problem was not only to describe how particles can exhibit wave properties (as is possible with Schrödinger's and Dirac's equations), but also to do the opposite, i.e., explain how waves can behave like particles. Or more specifically, explain how electromagnetic waves can become photons.

> Physicists sought a theory beyond quantum mechanics, in which not only could particles be waves, but waves could also be particles. Such a quantum theory of the physical fields (waves) was also required to explain the wave–particle duality.

© Springer Nature Switzerland AG 2018
L. Jaeger, *The Second Quantum Revolution*, https://doi.org/10.1007/978-3-319-98824-5_11

And there was yet another fundamental problem that physicists wanted to overcome. Since the early days of quantum physics, it had been known from experiments that there were processes in the microcosm in which particles simply disappeared or emerged out of nowhere. However, in the equations of Schrödinger and Heisenberg, there was no room for such processes; they described phenomena in which the number of particles always remains constant. They could not therefore consistently treat the interactions of matter and electromagnetic radiation, in which particles are both created and destroyed. The best known example of such an interaction is the absorption or emission of photons by atoms. But neither could the electron–positron pairs arising from the interpretation of Dirac's equation, and predicted to emerge spontaneously in the vacuum, be described by the quantum mechanics of Schrödinger and Heisenberg.

> By 1930, quantum theory was neither able to describe the absorption or emission of photons, nor the spontaneous appearance and disappearance of electron–positron pairs in the vacuum.

From the Classical Particle to the Quantum Field in Two Leaps

While so far most (thought) experiments had started from the notion of particles—just think of the double-slit experiment in which electrons or photons are fired as particles at the apertures—physicists now turned their attention to fields. A physical field is nothing more than the spatio-temporal distribution of a physical variable. Each position in space and time is assigned a value of the physical (field) quantity. These field variables can be vectors, i.e., with spatial orientations (such as electric and magnetic fields) or they can be non-directional quantities (such as the thermodynamic fields of air pressure, temperature, and density). The former are called vector fields, the latter scalar fields.

In order to arrive at a quantum field theory, physicists had to treat both types of field in a quantum physical way:

1. The electromagnetic field (a vector field) in order to describe the photon, and its absorption and emission when it interacts with atoms.
2. The wave function of quantum mechanics, which strictly speaking is nothing other than a scalar field: each point of space receives a value

whose square represents the probability of the particle's being found there (or, depending on the representation, the probability of another state variable). Although the field described by a particle wave function is a result of quantum mechanics, it describes continuous spatial and temporal field quantities. In this sense, quantum mechanics can be understood as classical field theory. In order to describe the formation and destruction of particles in the Dirac theory, this "classical field" had to be quantized.

> The decisive step in quantum field theory was to interpret, not only the electromagnetic field, but also the wave function of a particle as a quantum field.

It was Paul Dirac who in 1927 worked out the first steps to be taken to move toward such a description of quantum fields.[1] Physicists speak of "quantization of the field" or "second quantization".

- First quantization takes place in the arena of quantum mechanics. It replaces the classical physics of a particle with a wave equation, the Schrödinger equation, and a wave function, solution of the wave equation.

> First quantization is nothing other than the derivation of a wave function that transforms classical particle properties into wave properties.

- This wave function is then quantized again. Mathematically, this means that the field quantities themselves become operators. These are instructions for the user to perform certain calculations on given mathematical entities (e.g., numbers or functions). A very simple and well-known operator, for example, is the plus sign (called the addition operator), which adds a certain number to a given number. It is obvious that, when a *function* becomes an operator, the complexity of the calculations increases significantly.

For the mathematically advanced reader, it will pay to take a closer look at this essential step (those less interested can skip until after the next box).

---

[1] P. Dirac, *The quantum theory of the emission and absorption of radiation*, Proc. R. Soc. London A 114, 243–265 (1927).

The basis of second quantization is the quantum theoretical description of many-particle systems. In Schrödinger's representation, the wave functions of such systems are elements of a tensor product of one-particle Hilbert spaces. In a first step, these wave functions are transformed (mathematically this is equivalent to a base transformation) in such a way that, instead of representing individual particles in an ensemble, they now describe an integrated system in which the various possible states are characterized by which one-particle quantum states are occupied and which are not. The following will illustrate this transformation. In an infinite system of mailboxes, some of the boxes contain mail, others not. Instead of checking which box *a specific letter* has been posted in, we calculate which of the infinite number of mailboxes contain letters and which are empty. An essential condition for this procedure is the indistinguishability of particles. It ensures that the two descriptions, the one which looks at individual particles and describes them in the ensemble and the one that just counts the occupied states in the overall ensemble, are identical.

The new state space is now a space containing all states and their associated degree of occupation (their associated particle number). Physicists speak of the Fock space, named after the Russian physicist Vladimir Fock, who in 1932 introduced this key mathematical concept for second quantization.[2]

What follows is the actual quantization process. Every possible occupation state is now assigned two operators. For the given state, one of these creates a further occupation, i.e., a new particle in that state, the other erases an existing occupation.

> In second quantization, the associated one-particle wave function becomes an operator for each position, determining whether the state remains the same or a particle is either generated or destroyed in this state.

In an analogous way to the second quantization of the wave function described above, the classical field equations of electromagnetism can be treated quantum theoretically. In the resulting theory known as quantum electrodynamics, electromagnetic fields become quantum fields.

It is somewhat confusing that the quantum-physical treatments of both the wave function and also the electromagnetic field are referred to as

---

[2]W. Fock, *Konfigurationsraum und zweite Quantelung* (Configuration Space and Second Quantization), Zeitschrift für Physik **75**, 622–647 (1932).

"second quantization". Because in fact there is only ever a first quantization! Mathematicians do not have to perform such calculations twice in succession, the wave function or the classical electromagnetic field are only quantized once.

> The second quantization of the wave function and the electromagnetic field led to a new area of quantum physics, called *quantum electrodynamics* (QED).

The New World Takes the Lead

While the previous theory of quantum *mechanics* had proved intuitively inaccessible, difficult to understand, and mathematically inscrutable, physicists soon realized that they needed an even higher level of abstraction and far more complex mathematics for the development of quantum *field theories*. The quantization of fields required a completely new approach.

In the first half of the 20th century, the leading figures in theoretical physics came from German-speaking Europe: Planck, Einstein, Bohr (who was Danish and worked in German), Heisenberg, Pauli, Schrödinger, Sommerfeld, Ehrenfest, and Born had been the pioneers of quantum mechanics and relativity. But in the late 1940s, the European dominance of theoretical physics was broken. The formulation of quantum field theory was the work of a new generation of physicists, and these came more and more from a rapidly emerging country of scientific research: the United States of America.

The theories produced by the physicists of the Old World were the fruit of a long and intense reflection on abstract concepts such as space, time, matter, force, and motion, together with their concrete meaning. It was not until they had clarified these questions that theoretical physicists could use mathematics to set their theory in the right form. Modern quantum field theories, on the other hand, represented the triumph of a new way of doing theoretical physics. The style of the American Anglo-Saxon scientific tradition was very different from the European one. It was pragmatic and sober and put much more emphasis on mathematical virtuosity than on the ability to think deeply about difficult conceptual problems. Put simply, while the European tradition started with concepts and then set these in mathematical form, the Americans started with mathematics and then thought about the physical interpretation. "Shut up and calculate", was how the physicist

David Mermin summed up this new methodological style in theoretical physics.[3]

> Einstein and his colleagues saw their work as part of a broad philosophical tradition. The new Anglo-Saxon form of theoretical physics was much more abstract. Its homeland was mathematics.

For the quantization of classical fields, some difficult conceptual and mathematical hurdles had to be overcome and sophisticated ideas had to be developed. Initially, the formulation of quantum field theory was in a large part mathematically unsatisfactory. Only the more comprehensive framework provided by the Lagrangian formalism would yield a mathematically consistent picture. With it physicists finally succeeded in consistently transforming wave functions or field variables into operators.

As new and abstract as the underlying mathematics may sound, physicists had already become familiar with its basics from Heisenberg's formulation of quantum mechanics. There, in each individual location, the momentum or energy was replaced by an operator (or a "matrix"); in quantum field theory it is the wave function or the field quantities that are subjected to this treatment.

So in the new paradigm of quantum field theory, the bottom line was the following picture:

- In quantum mechanics, the state space of a physical object was the spatially non-located (albeit spatially describable) wave function.
- Quantum field theory no longer considered wave functions whose arguments were the possible *locations* of a particle. The centre of interest became state functions whose arguments were all possible *field configurations*. This perspective once again vastly increased the space of state functions.
- Each field configuration corresponded to an operator acting on an abstract multi-particle state space, the Fock space. In this space, states with different numbers of particles could now be considered and transformed into each other.

---

[3]N. David Mermin, *What's Wrong with this Pillow?* Physics Today, April 1989, p. 9. This phrase is often also attributed to Paul Dirac and Richard Feynman.

Quantum field theory was eventually able to describe all the quantum properties of physical fields, the effects of photon emission and absorption, and the creation and annihilation of electrons and positrons in electromagnetic fields, as predicted by Dirac's theory.

> The quantum field became the fundamental concept of physics, from which all properties of matter, fields, and forces could be derived.

As abstract as the new theory was, it did allow a degree of correspondence between theory and experiment that had never before been achieved in the history of science. The discrepancy between experimental measurements and theoretical calculations turned out to be in the range of 0.0000000001%. By comparison, using Newtonian physics, the orbit of the moon (which today can be determined to within a centimetre) could only be calculated with about 99.3% accuracy. If Einstein's general theory of relativity is taken into account, the difference between the calculated and the measured position of the moon becomes much smaller, but it does not come close to the accuracy of quantum field theory (which is, of course, also due to the fact that we do not know the precise mass distribution in the Earth and the Moon, as well as the gravitational effects of the other planets).

Let me just reassure readers who may have found the last few pages too mathematically abstract. Physicists were excited to have taken such a big step forward, but no one actually had much idea what a quantum field really was.

> Even though quantum fields could be calculated and the correspondence between these calculations and experiment was truly breath-taking, the true nature of the quantum field remained incomprehensible.

We will return to the question of the nature of the quantum field in Chap. 15 when discussing the philosophy of quantum theory.

Quanta of the Electromagnetic Field—The New Role of the Photon

Even if quantum fields elude any clear explanation, the quantization of electromagnetic fields bears great significance for physics:

- From quantum *mechanics* we derived the quantization of energy for particles in bound systems. For electrons bound in atoms, this means that they can only take in and release energy packets of specific sizes (although free electrons can assume any energy state, i.e., their states are not quantized).
- Quantum *field theory* showed that even the physical quantities associated with an electromagnetic wave can only assume quantized values. In any physical process, the energy of the wave is added or absorbed only in "packets"—and these packets are nothing other than photons. Quantum field theory thus provided the theoretical validation for the existence of the photons whose properties Einstein had first described in 1905.

Photons are the quantum packets that can be extracted from or inserted in an electromagnetic wave.

The true explosive power of quantum field theory is revealed when we consider, not only single quantum objects, but their interactions with each other. More specifically, the following picture emerged: an electron generates a photon out of its own electromagnetic field, and this is absorbed by another electron. This exchange transfers the electromagnetic force from the first to the second electron.

To illustrate the process of energy transfer in the quantum world, we can imagine the electromagnetic field as a stretched rubber sheet. If a massive ball (electron 1) is thrown onto it, its energy is transferred to the rubber sheet in the form of a wave. In the quantum world, this energy transfer occurs instantly and completely, the ball sticks to the sheet without swinging. The wave propagating across the rubber sheet causes a second ball at another location (electron 2) to be thrown up in the air, the necessary energy being provided by the wave in the sheet. In the quantum world, however, the balls on the sheet can absorb only waves of very specific quantized energies which make them jump. This presents no problem because, in the quantum world, resonance prevails—this is the ideal case where the energy of the wave and the properties of the balls match perfectly. The second ball completely absorbs the energy of the wave, and the wave thus disappears.

In the world of quantum mechanics, the *electron* still jumps from one state to the next when a photon excites the atom, but in quantum field theory, it is the *field* that gets excited and transformed into a state of higher energy.

Now comes the trick: these waves cannot be calculated as waves, but only as particles. They do not *act* like waves, but run from one ball to another, like a ball in a pool transmitting its energy. Physicists thus call these field quanta *exchange particles*.[4]

These particles, which manifest themselves out of an electromagnetic wave and transfer energy from one quantum object to another, are the photons. With the demonstration that photons were exchange particles, physicists had obtained two insights:

- Finally, nearly thirty years after Einstein had introduced the particle nature of light in the form of photons in 1905, it had now been explained theoretically and proved mathematically that electromagnetic waves could manifest themselves as particles.
- Electrically charged particles only attract or repel each other because photons, as exchange particles, transmit the electromagnetic force. Later, other exchange particles were discovered, which mediate other forces.

Interactions between quantum objects are manifested by the exchange of particles, so-called field quanta or exchange particles. Physicists believe these to be the basis of all physical forces.

Particles Out of Nothing

In general, field quanta can exist in real and in virtual states:

- In the virtual state, they do not appear concretely as particles or radiation, but instead, as exchange particles, representing the effects of a field. The virtual particles also include the electron–positron pairs spontaneously appearing in the vacuum, as implied by the Dirac equation and the Heisenberg uncertainty principle, and which immediately decay again (see Chap.10). Recall that such spontaneous fluctuations of a quantum field are not visible or directly measurable, but do have an indirectly measurable effect on physical quantities.
- Real, long-lived, and actually observable particle–antiparticle pairs can also arise spontaneously, although this requires a supply of energy. With

---

[4]There is still a photon field whose excitations are photons. These also have wave characteristics, as expressed at the double slit. In a moment, we will discuss the difference between *virtual* and *real* photons.

the quantization of the Dirac wave function and the electromagnetic field, it has even become apparent that virtual exchange particles can change into real states. These then move freely through space and can be detected as radiation.

As excitations of a quantum field like the photon, real particles are stable and observable. Of course, a photon of sufficiently high energy does not simply carry an electron–positron pair piggy-back, and then release when an opportunity arises. In fact, real matter arises from the energy of the photon, and this matter is itself a quantum particle that can manifest itself as a wave or particle.

> Through quantization, electromagnetic waves obtain particle characteristics. Photons are to be understood as a manifestation of the excitations of quantum fields.

These results provided the (theoretical) proof that, apart from disintegrating into energy, matter could also arise spontaneously from energy. This is the deeper content of Einstein's formula $E = mc^2$, i.e., the equivalence of energy and mass. In quantum field theory, we thus lose the last remnants of particles as entities with unchanging material properties.

So we have closed the circle. We have rediscovered the wave–particle duality, but this time from the other side. When physicists developed quantum mechanics, they had to adjust laboriously to the fact that particles behave like waves (such as the electron at the double slit), and now, under the spell of quantum field theory, they had to rethink everything again, conceiving of waves in the electromagnetic field as moving from one quantum object to another in the form of particles, and indeed calculating them as such.

> Quantum field theory states that fields and particles can emerge from each other. They are one and the same. This finally removes the last elements of any contrast between particles and waves.

Although it is clear that the terms "particle" and "wave" no longer have any place in quantum physics, these terms have proved to be surprisingly long-lived. This is probably because, in the abstract world of quantum physics, they serve as a last ditch attempt to maintain comprehensibility from the intuitive point of view of our perceptions. But one thing must be made clear: even if there is still separate talk of particles and waves, they are actually one and the same, that is, manifestations and excitations of quantum fields.

The Dematerialization of Matter

So how can we envisage the construction of our world? Quantum entities like the electron are inseparably linked not only to the measurement apparatus of the observing subject, as quantum mechanics already showed, but also to their own interactions. They are constantly surrounded by fluctuating quantum fields and a corresponding cloud of virtual particles of all kinds that are continually emitted and absorbed, and they are also surrounded by the interactions and the exchange particles related to these. Even in a vacuum, which seems to be empty space, virtual particles can emerge out of nowhere, and if the necessary energy is provided, real (observable) particles and their interactions can also emerge.

A quantum particle is thus no longer a very small piece of matter, as physicists from Democritus to Dalton, Thomson, and Einstein had once imagined, but rather an inseparable and interdependent cloud of fields, interactions, and other particles.

> The permanently arising and passing virtual particles and the interactions with other particles are essential components of the quantum entities themselves.

Classical physics and a significant part of the Western philosophical tradition had been based on the belief that it was the solidity of atoms that guaranteed the stability of matter. But with Bohr's atomic model it had already become clear that if 99.9% of matter is empty space, the solidity of objects like a table cannot come from the solidity of material atoms. So where does it come from then? Contrary to what we might imagine, it is the "immaterial" fields that give consistency and solidity to the material things of our everyday world; to be precise, it is the electromagnetic forces *between* the electrons and the atomic nucleus.

With quantum field theory, the last features of the classical idea of solid matter and the substantial integrity of material things finally dissolved. As we delve more deeply into the basic structure of matter, the familiar image of solid particles becomes more and more blurred, and the substance behind these phenomena seems to fade.

> The core message of quantum field theory literally turns our world upside-down: it is not the material essence of minute particles that keeps the world together at its very core, but the interactions between these particles.

Everything Is One

Could it not be that, if we take an even closer look, we will discover the tiniest material particles awaiting discovery at the *very* end of the scale? It seems likely that we will never know, because in this respect, our thirst for knowledge is confronted with a fundamental limit.

The smaller a structure, the more energy is required to observe it, e.g., in the form of light at ever shorter wavelengths. A classical light microscope cannot resolve structures below the wavelength of visible light (about 500 nm), and in practice, the limit is even more restrictive, as other parameters come into play. We can resolve smaller structures with higher-frequency electromagnetic radiation, for example X-rays. In general, in order to examine the properties of very small objects we must shoot particles of correspondingly high energy at them. A crude basic rule is that, for a spatial resolution of one femtometre (10^{-15} m, approximately the radius of a proton), an energy of about 200 GeV is required (1 giga electron volt $= 10^9$ eV $=$ one billion electron volts, where 1 eV is the kinetic energy of an electron when accelerated through a potential difference of one volt). Observation of smaller structures requires bombardment with correspondingly higher energies.

> In order to achieve ever finer resolution of spatial structures, higher and higher frequencies, i.e., higher and higher energies are required.

In practice this means that ever larger (and more expensive) machines need to be built to inspect ever smaller structures. The gigantic particle accelerator called Large Hadron Collider (LHC) at the CERN research centre in Geneva consists, among other things, of an circular underground tunnel with a diameter of over 36 km. Here, particles are provided with tremendous amounts of energy by accelerating them to within a miniscule fraction of the speed of light. The particle projectile acts like a hammer when it hits another particle, cracking open the target particle like a nut, so that its components become visible. But for very small structures, the probability of a hit is correspondingly low. In this context, physicists also speak of the "cross-section" of the particle to be investigated, where the cross-section is the area that the accelerated particles must hit in order to resolve its structure.

With energies up to 14 TeV (1 tera electronvolt $= 10^{12}$ eV), physicists can achieve a resolution of about 10^{-18} m (structures with cross sections of 10^{-36} square metres). This energy of 14 trillion electron volts "only" corresponds approximately to the kinetic energy of a fly. In the LHC, however, this energy is concentrated in an area that is about a thousand billion times smaller than a fly. Enough scope to get to the bottom of things, one might think.

However, in quantum field theory, below a certain scale, the energy used to achieve this spatial resolution will itself produce particles, among them exactly those we were trying to observe in the first place. Thus the desired decomposition of a quantum particle to determine its structure and properties turns into a process that creates new versions of it. The observing instrument gets transformed into the very thing that was to be observed.

> If the energy used for an observation exceeds a certain value, it will be transformed into the very particles that were to be observed in the first place. Therefore, beyond a certain point, we cannot look more deeply into quantum structures.

The interdependence of quantum objects and their interactions, and the permanent transformation of matter into energy and vice versa, finally provide us with an answer to the ancient question about the nature of the very smallest particles: they do not exist. They simply dissolve into the appropriate set of quantum entities and their interactions. Nevertheless, physicists simply assume that elementary particles are point-like, because with ever greater expenditure of energy, their spatial extent can be pushed below any limit reached so far without them showing any observable intrinsic properties.

> Quantum field theory answers one of mankind's oldest questions: it tells us that there is no lower boundary to the size of elementary particles. The deeper we peer into the quantum world, the more the concept of matter fades before our eyes.

12

Infinity Minus Infinity Gives Something Finite: How Physicists Learnt to Deal with Infinitely Large Values in the Infinitely Small

The quantum field theory of the electromagnetic field addressed yet another persistent problem of classical field theory: the retroactivity or reaction of the electron's radiation on itself. Already around 1900, physicists discovered that an electron that moves through an electromagnetic field is not only passively exposed to the forces of this field, but also radiates a field that acts back on itself. This led Max Abraham and Henri Poincaré to speak of the "electron's self-energy".

This energy inherent to the electron should influence another important physical property, namely its inertia, i.e., its resistance to changes in motion, and therefore its mass (higher inertia means nothing other than higher mass). Physicists therefore spoke of the "electromagnetic mass" of the electron, which should make a certain contribution to its total mass. Some physicists even suggested that there is no "bare" mass of the electron, rather that the entire mass of an electron may actually be its electromagnetic mass.

The electron self-energy thus affects the inertia of a particle, i.e., its mass—this was the first time a direct connection had become apparent between energy and mass. Some years before Einstein produced his famous formula, Abraham and Poincaré had already determined the value of the electromagnetic mass to be $m_{em} = E_{em}/c^2$. At first, Einstein himself had also limited his statement that the mass of a particle increases with its energy to the case where the extra energy comes in the form of electromagnetic radiation. However, he was the first to recognize the universality of the equivalence of energy and mass, expressed in the formula $E = mc^2$.

© Springer Nature Switzerland AG 2018
L. Jaeger, *The Second Quantum Revolution*, https://doi.org/10.1007/978-3-319-98824-5_12

> The physically observable mass of an electron consists of two components, a bare constituent and a part which arises through its own field, i.e., its self-energy.

Open Sesame!

Of course, the field of the electron not only has an effect on its own mass. It also interacts with the field in which it is traveling. If this effect is taken into account, the standard formula for the (Lorentz) force acting on an electron in an electromagnetic field is augmented by another term and the Lorentz equation becomes the Abraham–Lorentz equation.

However, this beautifully derived equation from classical field theory comes with a fundamental mathematical problem: if the electron, while possessing electric charge and mass, occupies no space, as physicists suppose, then at ever smaller distances the electromagnetic forces that act upon it, and hence also its electromagnetic mass, take on higher and higher, and eventually infinitely high values. How can this be possible?

Physicists got themselves out of trouble by using a trick. They said that if certain variables in the Abraham–Lorentz equation take infinitely high values, then others, like the one describing the (unobservable) bare mass of the electron, would have to take on a value such that the *sum* of the two terms yields a finite value. In other words, they assumed that the bare mass of the electron is also infinite, but with a negative sign—infinity minus infinity can yield a finite value again. This little piece of trickery can never be refuted because it is impossible to determine the bare mass of an electron separately from the mass produced by its charge. Only the sum of the two terms can be measured.

> By a clever trick, physicists managed to retrieve the mass of an electron from the realm of the infinite. They simply stated that the bare mass of the electron is "minus infinity."

But even with this interpretation of a "renormalized" mass, the problems of the radiation reaction of the electron did not completely vanish. The infinities proved quite persistent, because the new term in the Abraham–Lorentz equation made possible an acceleration, which could sometimes lead to infinitely high velocities. It was like dealing with a tablecloth that is too small—one has barely arranged it to cover one half of the table and one sees that it is not enough to cover the other half. No matter what physicists came up with in classical electrodynamics, they were unable to describe the radiation reaction of the electron theoretically without contradiction.

They ultimately came to realize that the problem of infinities kept cropping up again and again in their equations, in the least-expected places, because an age-old question remained unanswered: what is the world like on the smallest length scales? Democritus and Kant had already reflected upon this problem. It was already known in the early twentieth century that an atom could not just be a massive ball. It consisted of protons, neutrons, and electrons—with a great deal of nothingness in-between. Looking deeper into smaller and smaller dimensions had not brought much of an answer. The question of the innermost structure of the smallest particles had basically just focused on a new object: instead of the atom, the electron and proton now stood at the gate whose key was missing.

> In the infinities that appear in classical field theories, we come up against the ancient question about the smallest structure of matter.

In the previous chapters we heard again and again that quantum mechanics or quantum field theory were expected to remove many problems that had annoyed physicists and philosophers for a long time. But in the case of the ever-recurring infinity, this would take much longer than hoped. Even in the context of quantum electrodynamics, it proved hard to get rid of them at first.

For example, these infinities unexpectedly arose in 1948 when Richard Feynman, one of the founders of quantum field theory, developed an ingeniously simple, intuitive, and at the same time computationally complex way to present effects in quantum field theories—the Feynman diagrams.

Spirals and Loops

In classical physics, particles have a well-defined position at all times. If the particle moves, it has a trajectory. In mechanics the interactions between trajectories are described by collision processes, in classical field theory by field forces.

The description of the interactions between quantum objects is fundamentally different. There are no longer particles with defined trajectories interacting directly with each other; the idea of electrons with a circular orbit moving around the atomic nucleus has thus become obsolete. Rather, quantum objects are described by wave functions that correspond to widely scattered trajectories Physicists therefore discuss the theoretical description of interactions in quantum fields under the umbrella of "scattering theory", which describes the dynamics of scattering of waves and particles mathematically.

In the quantum world, there are no fixed trajectories of particles, as they are described and calculated in classical physics, but very many, widely distributed *possibilities* of trajectories.

The goal in describing the dynamics of interacting quantum objects is to calculate the probabilities of all transitions of states of the system *before* the interaction to possible states *after* the interaction (the transformation being given by the so-called S-matrix). This requires the calculation of many complicated integrals over many variables, and involving many different contributions.

Feynman, however, took advantage of the fact that the integrals of the S-matrix have a very regular structure which can be traced back to elementary mathematical building blocks. His main trick was to introduce certain mathematical operators, called propagators, for a theoretical description of the behaviour of the wave function during the interaction. He drew diagrams whose lines represent concrete computational rules and thus help theoretical physicists to keep track of these complex and lengthy calculations.

Feynman diagrams illustrate elementary processes in quantum fields. They look simple and clear, but the underlying mathematics and computational requirements are very complex.

Here are some examples of rules for the Feynman graphs, each of which stands for certain fixed mathematical terms:

- The propagator of the electron and other elementary particles is assigned a continuous straight line.

- Particles exchanged in the interacting quantum field, such as the photon for the electromagnetic field, are represented by wavy or spiral lines.
- Vertices are the junctions of two lines where a propagator and a exchange particle meet.
- Lines connected to a vertex only at one end are considered to be real particles, while lines connecting two vertices represent virtual particles.
- Closed lines or loops of virtual particles may occur. The physical phenomenon behind these is the spontaneous generation of a (virtual) particle–antiparticle pair. For example, an electron–positron pair can emerge out of a photon and immediately disappear again.
- A loop also appears in the Feynman diagram when an electron emits a photon and immediately absorbs it again. This is the quantum field theoretical description of the above-mentioned interaction of the electron with itself. In the electromagnetic field of the electron, photons constantly arise and disappear.

Of course, Feynman's propagators do not depict any real paths of particles or locations of interaction. They should not be understood as a description of actual spatiotemporal processes. The same applies to the term "exchange". The mediation of an interaction between two quantum objects by exchange particles is only an attempt to make a process that occurs outside of any spatiotemporal frame comprehensible to the human mind.

> Feynman graphs are read as calculational instructions and do not map actual spatiotemporal events. Moreover, the idea of exchange particles is just an auxiliary construct for the purposes of illustration.

By the way, Feynman was not only a great physicist (he was one of the greatest in the second half of the 20th century), but also an exceptionally good teacher. His textbooks are still very popular with students of theoretical physics. In addition, he was also an amusing writer. His autobiography "Surely you're joking, Mr. Feynman!" is highly readable and has become a bestseller.

The utility of the Feynman diagrams can be seen, among other things, from the fact that, with their help, the g-factor of the electron mentioned in Chap. 10 can be calculated to an extraordinary level of accuracy. The theoretical value which results from quantum electrodynamics and the corresponding Feynman graphs, agrees with today's experimental measurements up to 12 decimal places, a hitherto unsurpassed correspondence between theoretical calculation and experimental measurements.

The accurate calculation of the g-factor gave the new theory of quantum electrodynamics a certain level of credibility. This was strengthened when it turned out that, in parallel and independently of Feynman, two other physicists had developed the theory of quantum electrodynamics: the American Julian Seymour Schwinger and the Japanese physicist Shin'ichirō Tomonaga. During the Second World War, communication between Japanese and American physicists was interrupted, and the various approaches could only be standardized after 1945 to form *one* coherent theory. Feynman, Schwinger, and Tomonaga shared the 1965 Nobel Prize in Physics for their joint development of QED.

Complicated Without Limits

So let us go back to the infinities. With the advances in quantum mechanics, the classical problem of pointlike particles and their infinitely large mass and energy seemed to have been resolved. For if the charge of the electron has a fuzzy location due to the Heisenberg uncertainty principle, it cannot be considered pointlike. But no sooner had physicists considered the problem solved than the infinities sprang up again in another place, in fact, in the Feynman diagrams. Many of the integrals represented by loops in Feynman diagrams have no finite value. In mathematical terms, they diverge. This in turn yields infinite values for the electric charge and the mass of the electron.

Naturally, infinitely high values for mass, energy, and other physical variables are, in physical terms, just as impossible in quantum field theory as they were in classical field theory. Even the brilliant Dirac gave up in desperation trying to solve the problems of infinities as they show up in these diverging integrals.

> The problem of infinities also occurs in quantum field theory. Loops in Feynman diagrams with virtual particles lead to infinite contributions to the S-matrix.

Intuitively, the reappearing infinities can be explained as follows. When an electron moves in a quantum field, it permanently emits virtual photons which it absorbs again shortly afterwards. Each photon in the cloud that the electron carries along with it can in turn generate virtual particle–antiparticle pairs (electron–positron pairs, but also other pairs). And out of these, in turn, virtual photons can emerge again, and so on. So around each electron there is a cloud of countless virtual particles, and there are endless possibilities for the behavior of that virtual cloud. According to Feynman's rules of calculation, we have to add up (integrate over) *all* these possibilities, and this eventually leads to infinite values for the components of the S-matrix.

But in the end, theorists did find a way to deal with these infinities. Feynman and others developed a technique that their predecessors had already used in classical physics: renormalization. With this, terms with negative infinite values are pragmatically added to the terms with positive infinite values, in such a way that the sum of the two turns out finite.

Beyond the Horizon

Another way of dealing physically with the infinities is to interpret the prevailing (non-renormalized) quantum field theory as an effective theory, valid only within a certain energy range. While this energy range covers everything that has so far been achieved experimentally (and probably quite a bit beyond that), it is, at any rate, finite in terms of the energies considered within it. In the mathematical processing of Feynman diagrams, however, we must integrate over *all* energy ranges. In describing the path of an electron through a quantum field, we must also consider those energies that are as large as that of our entire galaxy. No wonder these integrals provide nonsensical (infinite) results.

> One reason for the unwanted infinities in the Feynman diagrams is that, according to quantum field theory, energies that are likely to exceed its own validity must be taken into account.

Once again, the proven method of renormalization helps us to handle the infinite energies. The way out is to perform the calculations only up to a certain energy. When calculating Feynman diagrams, we are faced with the task of arbitrarily selecting the energy scale up to which we wish to consider contributions, i.e., fixing what is known as a *cutoff*. This exercise is referred to as regularization of the energy scale. With such a constraint the integrals at last yield finite values. Depending on the selected cutoff, however, the calculations provide different values. Only after a corresponding renormalization of the particle parameters (which depends directly on the chosen cutoff and not on any desire to get as close values as possible to the measured with our calculations), do these differences balance out as if by magic. Contributions that originate from energies beyond the cutoff, and which would come into the calculations if we chose higher cutoffs, can be compensated with the appropriate renormalization, whatever their value. Theoretically calculated and experimentally measured values are thus in perfect agreement.

> Only with a *cutoff* and a corresponding renormalization can infinite values be avoided in the calculations relating to events in the quantum world. In the end, physical processes can thus be treated in a consistent manner.

But isn't this just an enormous bamboozle? An infinity is removed by adding another infinity, or the considered range of possibilities is arbitrarily curtailed, and then, like a rabbit out of a magician's hat, the desired finite value for the relevant physical quantity precisely matches its experimentally measured value. But there is one difference with previous methods and theories: in contrast to renormalization in classical physics, it leads in quantum electrodynamics to consistent results, because we can always find counter-terms in the scattering matrix which, upon appropriate interpretation as additional contributions to physical quantities such as masses or charges, eventually make all the infinities disappear.

> For physically measurable quantities, renormalization solves the problem of infinite terms in the Feynman diagrams.

Remaining Discomfort

Of course, not all physicists are comfortable with this process. Most physicists admit that they are simply sweeping the infinities under the rug. Dirac remained a critic of renormalization throughout his life. And even Feynman said:

> But no matter how clever the word [renormalization], it is what I would call a dippy process! Having to resort to such hocus-pocus has prevented us from proving that the theory of quantum electrodynamics is mathematically self-consistent. It's surprising that the theory still hasn't been proved self-consistent

one way or the other by now; I suspect that renormalization is not mathematically legitimate.[1]

Renormalization theory is currently the last chapter in physics when it comes to dealing with the smallest structures in nature. But physicists have not yet reached this point: for a quantum theory of gravitational forces, which faces the same problems, renormalization does not work. Here, an infinite number of counter-terms is needed to eliminate all infinities. For this reason, the different contributions for different cutoffs do not balance out. Physicists say that a quantum version of gravitational theory is not renormalizable. We return to this problem in Chap. 14.

> While some subtle mathematical tricks have pushed the boundaries of computational predictability to ever smaller dimensions, the ultimate problem of infinity has not yet been solved.

[1]Feynman, R, *QED, The Strange Theory of Light and Matter*, London, 1990, p. 128.

13

More and More Particles: From the Particle Zoo to the Standard Model of Elementary Particle Physics

In general, physicists are considered rather sober individuals. Only in exceptional situations can we observe collective emotional outbursts within their ranks. On July 4, 2012 such a moment had come! On that day, the European Nuclear Research Center CERN, headquarters of the largest and most powerful particle accelerator in the world, the *Large Hadron Collider* (LHC), announced that it had detected the Higgs boson, a particle that physicists had been seeking for decades as the last missing link in the Standard Model of particle physics. The champagne corks popped, and for once, physics dominated the headlines of the global press.

In public, the Higgs boson is sometimes also called the *God Particle*. This name goes back to physics Nobel laureate Leon Lederman, who had actually wanted to call it the "goddamn particle" because it was so difficult to detect. The particle had already been postulated as the product of an extremely abstract mathematical version of quantum theory, way back in the 1960s. But within this theory, it has a special, very *concrete* meaning: only the Higgs boson can give matter any mass. The Standard Model of elementary particle physics, including the Higgs boson, represents the current state-of-the-art in quantum theory, and it is the subject of the present chapter.

Confusion in the Particle Zoo

With Dirac's anti-electrons, a new branch of physics was born: particle physics. Physicists had realized that the world consists of more than just protons, neutrons, and electrons. The positron was predicted theoretically

© Springer Nature Switzerland AG 2018
L. Jaeger, *The Second Quantum Revolution*, https://doi.org/10.1007/978-3-319-98824-5_13

before it showed up in the laboratories of experimental physicists. But in the next thirty to forty years the order of things was reversed: many unknown and unexpected particles were found in experiments, and theoreticians were faced with the task of explaining them. This began in the years after World War II, when physicists began to look more closely at cosmic radiation and discovered more and more particles with ever more exotic properties. Here are some examples:

- First there was the muon, a particle similar to the electron, but with a mass of about 200 times the electron mass (1937).
- In 1947, there was the π-meson, which the Japanese physicist Hideki Yukawa incorrectly declared to be the exchange particle mediating the strong interaction within the atomic nucleus.
- In 1949, physicists observed K^+ mesons created by the decay of the π meson.
- In 1951, two particles were discovered that left a V-shaped trace in detectors. There had to be an electrically neutral particle that decayed into two charged particles. These particles were called Λ^0 und K^0.

People began to ask why there were so many different particles in nature? How were they related to each other? Before they could even get the slightest hint of an answer to these questions, theorists found even more work loaded onto their plates, with the invention of a new kind of experiment which increased the flood of hitherto unimagined particles to new levels. In particle accelerators, particles are brought to very high speeds with the help of electromagnetic fields and then made to collide with each other. The higher the energy used, the more "fragments" are created, and the deeper physicists can penetrate into the structure of matter. The largest and best known of these particle accelerators today is located near Geneva, at the *Conseil Européen pour la Recherche Nucléaire* (CERN): the above-mentioned *Large Hadron Collider*.

From the 1950s, particle physics got into a state of ever greater chaos due to the constantly increasing numbers of newly discovered elementary particles.

New particles were attributed exotic names such as Σ-, Λ-, Ξ-, and Ω-hyperons (sigma, lambda, xi, and omega hyperons). The question of the mechanisms underlying their formation and decay, and also how they should be classified and described theoretically, was a great puzzle. At first, the experiments designed to capture the diverse properties of the elementary particles and to develop a coherent theory only created more confusion. But one thing was clear: the multiplicity of particles could only be investigated if more was known about the forces that held protons and neutrons together in the nucleus of the atom.

> One obstacle on the path towards a classification of the denizens of the particle zoo was the development of a suitable quantum field theory for the forces at work in the atomic nucleus.

The Noble Eightfold Path in Physics

Since the 1930s, apart from the well known electromagnetic force, physicists had been aware of two other forces at work in the atomic nucleus:

- The strong nuclear force that holds together the components of the atomic nucleus against the electromagnetic force, which tends to push protons apart.
- The comparably weak nuclear force which is responsible for a particular type of radioactive decay of atomic nuclei, the so-called beta decay, in which electrons or positrons are released.

Not only the many types of particles, but even the forces themselves lacked a fundamental theoretical basis and explanation. The long quest for a theory that would sort out the particle zoo and provide a proper description of these two forces was first undertaken in the 1960s with the reflections of another 20th-century genius, the American physicist Murray Gell-Mann.

Gell-Mann tackled the problem pragmatically. He first turned his attention to those particles which, according to experimental measurements, were subject to the strong interaction, the so-called hadrons (electrons, for example, are not included), and divided them into different groups. He observed that some hadrons barely differ in their properties. The proton and the neutron, for example, have almost the same mass and the same spin, and they are both subject equally to the strong nuclear force. They differ only in their charge. Furthermore, in radioactive beta decay, a neutron turns into a proton or vice versa (there are two different types of this decay), sending out a positron or an electron, respectively. The proton and the neutron had to belong to the same group, Gell-Mann concluded.

The sorting of the particle zoo looked like a jigsaw puzzle, comparable to the one Dmitri Mendeleev had solved about 100 years earlier when he began to classify the chemical elements according to his periodic table.

On the basis of their masses, most hadrons could be roughly divided into a scheme with two groups:

- Baryons, which include among other things, the proton and the neutron.
- Mesons, which are generally lighter particles.

Gell-Mann then used a mathematical concept developed by the 19th century Norwegian mathematics Marius Sophus Lie and which physicists had recently rediscovered for their purposes. In mathematical terms, this is the theory of continuously differentiable groups. How can we illustrate such "Lie groups"? Consider a sphere and all possible rotations about arbitrary axes through the centre of the sphere. These rotations taken together constitute a Lie group, which at the same time spans a three-dimensional vector space (one angle dimension and two axis dimensions/parameters). Since the sphere itself always remains unchanged upon such rotations, we also speak of the "symmetry group of the sphere" (mathematicians call it the SO(3) Lie group).

One of these groups conceptualized by Lie proved to be perfectly suited for Gell-Mann's purpose: it was a group bearing the mathematical name SU(3), the "special unitary group of complex rotations in the three-dimensional complex space". With it, the various hadrons in Gell-Mann's scheme could be wonderfully classified. It turned out that, within the scheme of the SU(3) group, eight members fit together—so there had to be eight baryons and eight mesons in that scheme (more groups were added later). The theoretical reason for this is that the SU(3) group is eight-dimensional, but it was only later that Gell-Mann realized that. However, he sensed that his groups of eight offered a basis for a possible classification of all elementary particles interacting via the strong nuclear force. Gell-Mann had a special sense of humour: he chose the name "eightfold path" for his classification. In Buddhism this describes the noble path towards the highest knowledge.

With Gell-Mann's classification of the eightfold path in 1961, a pattern began to emerge in the jungle of particles.

Three Quarks for Muster Mark

Like so much in quantum theory, Gell-Mann's classification scheme had at first been completely ad hoc. However, he was convinced that behind the eightfold path there lay a fundamental principle. But what was it? He came up with the idea that every hadron might consist of even smaller particles. He assumed that all baryons were composed of three of these mini-elementary particles and that all mesons were composed of two of them.

The naming of these mini-particles revealed Gell-Mann's fondness for strange names. He remembered a line from James Joyce's novel *Finnegans Wake*, which used the expression *Three Quarks for Muster Mark*. And so Gell-Mann called his hypothetical particles "quarks". In order to explain their symmetry and interactions within the context of the strong nuclear force and to be able to classify the hadrons properly, he had to assign another quantum physical quantity to these quarks, called *isospin*.

> Just as Pauli had assigned a spin to the electrons in an ad hoc manner, Gell-Mann provided the quarks with yet another quantum physical variable, called isospin.

Taking the isospin into account, Gell-Mann called the new hypothetical particles the up-quark (isospin pointing up) and the down-quark (isospin pointing down). According to his line thought, protons had to consist of two up-quarks and one down-quark, while the neutron had to consist of two down-quarks and one up-quark.

But besides protons and neutrons, there were all the other particles, each with its own special properties. Gell-Mann's attention shifted quickly to the so-called K-mesons, also called kaons. These are generated by means of the *strong* nuclear force, and it was therefore widely assumed that this force must also be involved in their decay. However, it turned out that kaons decay under the influence of the *weak* nuclear force. It was strange, then, that particles formed by means of the strong nuclear force would not also decay through it. For this reason, physicists referred to kaons and related particles (like the Λ particle, and later the Σ, Ξ, and Ω particles) as *strange particles*.

Gell-Mann concluded that another type of quark, the *strange quark*, should be connected to these strange particles. And for reasons of symmetry, this was joined by another quark which physicists called the *charm quark*. In 1974, a particle was indeed discovered that contained this new type of quark (the Ψ-meson). Over time, when even more particles were discovered, two additional quarks were added to the list: the *bottom* and the *top quark* (only discovered in 1984 and 1995, respectively).

> The quarks which bind in threes to form the previously known baryons and in quark–antiquark pairs to form mesons were soon joined by other quark types: strange quarks, charm quarks, bottom quarks, and top quarks.

Today physicists know that, with these new quarks and their corresponding isospins, there are many other baryon associations besides the original octet identified by Gell-Mann, e.g., a decuplet associating ten baryons, each with isospin 3/2.

In 1969, Gell-Mann received the Nobel Prize in Physics for his work. But he had not been alone in his discoveries. Independently of him, the Russian–American physicist Georg Zweig had discovered the classification scheme for hadrons and also postulated the existence of quarks (which he called "aces"). However, as far as the Nobel Prize was concerned, Zweig came out empty-handed.

Gell-Mann's scheme impressively captured all particles discovered in accelerators. Now the relationship between particle prediction and particle discovery turned back in favour of the theoreticians. We may give two examples:

- Gell-Mann's scheme initially lacked a particular baryon. He postulated the existence of this particle and predicted its properties. Shortly afterwards, in 1964, the Ω (omega) particle was detected, corresponding to the gap in his system.
- The meson octet also lacked an eighth meson. But in 1961, in the same year as Gell-Mann published his classification scheme, the appropriate η (eta) particle was found.

To date, many other theoretically predicted particles have been found in accelerator experiments, fitting perfectly into Gell-Mann's grouping scheme.

> Today physicists know about 150 baryons and 200 mesons, all of which have a well-defined place in Gell-Mann's scheme.

Imprisoned Quarks

However, one thing still remained problematic: no single quarks had ever been observed. Eventually, an explanation was found: quarks can only exist in packs of two or three.

With the classification of the elementary particles in the eightfold path, the quarks displayed another pattern: each of them can be assigned a property that determines the strength of its interaction with the strong nuclear force, just as the electric charge does for the electromagnetic force. But while the electric charge knows only two values—plus or minus—there are *six* values for the strong nuclear force. These quark states were attributed the names of six different colours, conventionally chosen to be red, blue, and green for the quarks, together with the corresponding complementary colours anti-red, anti-blue, and anti-green for the anti-quarks.

> Just as the electromagnetic force between two particles is the result of their electric charges, the strong nuclear force is a result of the so-called colour charge of the quarks.

Of course, these colour properties are not literally the red, blue, and green we know and love in everyday life. The names were introduced to make it possible for the human mind to separate possible from impossible combinations. For a combination of quarks can only exist if the resulting triplet or doublet is colour-neutral. There are two ways to do this:

- For hadrons consisting of three quarks, that is, the baryons, the following rule applies: the quarks must be one red, one blue, and one green, mixed together to give the neutral white colour, or again one anti-red, one anti-blue, and one anti-green, which together also yield white. These are the only viable combinations.
- Hadrons made of a quark and an antiquark, that is, mesons, comprise one green and one anti-green, one blue and one anti-blue, or one red and one anti-red. These combinations also "neutralize" the colours to white.

> The colour properties of quarks give the "theory of strong nuclear forces" its name: quantum chromodynamics (QCD). After quantum electrodynamics (QED), QCD is the second quantum field theory of elementary particle physics.

Quarks are subject to yet another special feature: within baryons and mesons, the farther apart the individual quarks, the stronger the nuclear force between them. This is reminiscent of a rubber band or a spring whose restoring force is stronger the further it is stretched out. If a quark moves away from its partner, the force with which it is retrieved automatically increases. So every quark is inseparably connected with its partner quarks within the baryons or mesons. As colour-neutral particles, they are shielded against the strong forces of other particles, for example in neighbouring atomic nuclei. This is why the strong nuclear force has such a short range.

Non-physicists may find the colour theory and quarks a bit playful. But the underlying mathematics is highly complex and at the same time beautiful in its structure and symmetry.

The Standard Model of Elementary Particle Physics

Physicists today distinguish four basic forces (or fields) in nature:

1. Gravity described by Einstein's general theory of relativity (see next chapter).
2. The electromagnetic force: the corresponding theory is quantum electro-dynamics, which describes the interaction of charged particles.
3. The strong nuclear force: the corresponding theory is quantum chromo-dynamics, describing the interaction of hadrons and quarks.
4. The weak nuclear force: this is described by the quantum field theory of the weak force (see below).

In the late 1960s, physicists Steven Weinberg, Sheldon Glashow, and Abdus Salam succeeded in presenting the quantum field theory of the weak nuclear force and quantum electrodynamics as two sides of a single theory. Today physicists speak of the "theory of the electroweak force". It describes the interaction of all particles that do not interact with the strong nuclear force, the so-called *leptons* (Greek for "light particles").

For three of the four fundamental forces, there exists a quantum field theory that describes the interactions between the relevant particles.

In connection with these three forces, physicists distinguish two types of elementary particle:

- the two times three quarks as constituents of hadrons, and
- the set of leptons, which also includes two times three particles: the electron and its associated (electron) neutrino, along with two similar but heavier particles, the muon and the tau particle, each with their own neutrino, the muon neutrino and the tau neutrino.

Quarks interact through the strong nuclear force and these interactions can be described by quantum chromodynamics (but note that they also carry electric charge). Leptons, on the other hand, are immune to the strong nuclear force. They only interact through the electroweak force.

In each of these quantum field theories, the interactions of the particles with fields are described by the corresponding field quanta. These determine the quantum properties of the given field and, as exchange particles, transfer its forces from one particle to another. They are collectively called "gauge bosons".

- In the case of quantum chromodynamics (strong nuclear force), the exchange particles are responsible for ensuring that the quarks remain within the hadrons and that atomic nuclei do not fall apart. Physicists therefore refer to them as *gluons*. Due to the eight-dimensional nature of the SU(3) Lie group of quantum chromodynamics, there exist eight different gluons.

Gluons have a special property that ensures that the strong nuclear force between quarks is so powerful: because they themselves carry colour properties, they themselves feel the forces they transmit. This means that the forces mediated by the gluons and holding the quarks together are tremendously strong (albeit only over a very short range).

In contrast, when photons travel through an electric field, they are not affected by the field themselves, because they carry no charge. Photons do not therefore interact with other photons. The situation is much more complicated in the case of gluons. Even if quantum chromodynamics is well established as a theory, it is tough work for theoretical physicists to calculate the enormously complicated terms in the corresponding equations. They have been calculating many of them for 50 years now.

The interactions of quarks and gluons and their many possible states are not yet fully understood. One example is given by the processes in a quark–gluon plasma. This is a state of matter that arises at ultra-high energies, in which the two types of particle fly around interacting violently with each other.

- For the weak nuclear force, physicists postulated the existence of three different field quanta (their underlying group, the SU (2) Lie group, is three-dimensional): the negatively charged W^- particle, the positively charged W^+ particle, and the electrically neutral Z^0 particle. They too interact with each other.
- The fact that the electromagnetic force knows only one field quantum, the photon, comes about because its Lie group, the U(1) group, is one-dimensional.

The whole set of elementary particles, viz., the six quarks, the six leptons, and all the exchange particles, were brought together in the 1970s into a unified theory that has since become known as the *standard theory of elementary particle physics*, or simply the "Standard Model."

> The *Standard Model* is the current state of the art in our understanding of the physical world. However, it does not include gravity.

In experimental terms, the Standard Model has been a great success story. There is to date not a single clear experimental finding that cannot be reconciled with it (although there are indications of phenomena that physicists might not be able to reconcile with it, e.g., dark mattter). It celebrated two of its greatest successes in 1982, when the postulated W^-, W^+, and Z^0 exchange particles of the weak interaction were discovered, and in 1995 with the detection of the last quark (the top quark).

The Particle Without Which Nothing Works

However, for a long time physicists were unable to detect one final particle in their experiments, whose existence is absolutely necessary to the Standard Model, and without which the entire theory would collapse: the above-mentioned Higgs particle. Only with its help were physicists able to solve a fundamental problem for the existing quantum field theories: how do particles obtain their mass?

According to a symmetry property of quantum field theories, elementary particles should actually have no mass at all. A specific external field, the so-called Higgs field, is responsible for endowing elementary particles with their observed masses. Through a complicated and exotic-looking mechanism formulated by the British theoretical physicist Peter Higgs in 1964, the Higgs field causes the electroweak force to break down into the weak and electromagnetic forces on a particular energy scale. Physicists call this mechanism "spontaneous symmetry breaking" (we shall deal with symmetry principles in modern physics in more detail in Chap. 18).

> In the Higgs field, an elementary particle is slowed down like a ball in a viscous liquid. This deceleration results in the particles having an inertia, and this amounts precisely to the property of having a mass.

Experimentally, the Higgs field manifests itself in a corresponding particle, the "goddamn Higgs particle" mentioned in the first section of this chapter. Physicists had searched for it for almost 50 years. Its discovery on July 4, 2012 was a historic moment for physics and the greatest triumph of the Standard Model.

Not yet the End

But even after the discovery of the Higgs particle, the Standard Model continued to suffer from at least two other fundamental theoretical problems (and yet another, as the next chapter will show):

1. The Standard Model is anything but clear and simple. Physicists consider a theory as "simple" if it is based on a single basic structure. In the case of elementary particle physics within a quantum field theory of all forces and particles, this would be a *single Lie group*. Furthermore, the values of its parameters should if possible come out of the theory itself. In contrast, the Standard Model incorporates two different basic models, i.e., Lie groups, namely the SU(3) group of quantum chromodynamics and the combined SU(2) and U (1) group of the electroweak theory. In addition, there are 19 free parameters that have to be determined experimentally, such as the masses of the quarks and leptons, the electric charge, and the strengths of the three fundamental forces (the so-called coupling constants).

> Many physicists dislike the complex and heterogeneous structure of the Standard Model.

2. Secondly, the Standard Model does not include the gravitational force. This force plays a special role in the quartet of the fundamental forces of nature. The other three—the electromagnetic, the strong, and the weak force in the atomic nucleus—can be described in terms of a quantum field theory. At the same time, it is precisely these forces that act at the level of atoms, while the comparably very weak gravitational force has no role to play in this microcosm. It is only when we consider the enormous masses of stars and planets that gravity plays its dominant role in the universe, and indeed in our everyday lives.[1]

As already discussed in the last chapter, the theory of gravity (general relativity) is not renormalizable. This provided a first indication that the Standard Model and gravitational theory are incompatible. And things get even worse: the theory of gravity in the form of Einstein's field theory of general relativity is fundamentally irreconcilable with the structure of any quantum field theory. The reasons for this will be the topic of the next chapter.

> For very basic reasons, gravitation cannot be set in the form of a quantum field theory. Without a common denominator, it seems impossible to unite the two theories into a theory of all matter and its interactions.

There are many reasons to think that the Standard Model is not the last word as a fundamental theory of all natural forces. Many physicists hold a deep belief that nature is basically very simple, and that there must therefore be an even deeper, more fundamental theory of the micro (and macro) world. They thus continue to look for an explanation of the world in which all complications dissolve and only pure simplicity and beauty remain.

---

[1]The fact that gravity, with all its relative weakness at short distances, acts effectively more strongly than the electromagnetic force over large distances is due to the fact that there exist no negative masses. The electromagnetic force is attractive and repulsive, depending on the sign of the charges. Thus, over long ranges, isolated charges are shielded by other charges, e.g., dipole clouds, which make the forces acting on them effectively shorter range.

14

Einstein Does Not Fit: The Fundamental Problem in Physics Today

The world view of modern physics is based on two fundamental theories: quantum field theory and general relativity. One describes the atomic and subatomic world of the microcosm, the other the macrocosm of galaxies and the universe as a whole. There is apparently no overlap between these theories of the extremely large and the extremely small. Both theories lose their validity or applicability as soon as they approach the scale appropriate to their counterpart. The gap between the two worlds is also called the mesocosm—the world in-between. Here apply the rules that describe the everyday world of us humans: the rules of classical physics.

Thus classical physics is a limit in two respects. On the one hand, it constitutes the limit of quantum theory when we move from the scale of the atomic world to that of larger systems. But it also proves useful in the description of planetary and galactic events when the mass concentrations or energy densities are too low for the theory of relativity to be applicable. Thus, classical physics is the link between quantum field theory and the general theory of relativity—and we humans live in the world where these two theories *almost* meet.

The extremely precise empirical and experimental confirmations of quantum field theory and the general theory of relativity are a triumph, but at the same time a problem for modern physics, because the two theories are perfectly incompatible.

© Springer Nature Switzerland AG 2018
L. Jaeger, *The Second Quantum Revolution*, https://doi.org/10.1007/978-3-319-98824-5_14

The reason for the impossibility of bringing general relativity and quantum theory together to form a unified world theory lies in their completely different concepts of space and time:

- In quantum physics, physical events are embedded in an internal time sequence and an independently existing external space, just as we intuitively perceive in our everyday world. Even for the special theory of relativity (which was painstakingly and at the same time elegantly integrated into quantum theory), space and time behave *statically*, despite being linked into an integrated space-time. Quantum theory and the special theory of relativity are therefore "background-independent theories", or put another way, particles and fields live in a static space-time background.
- In general relativity, on the other hand, time is not an external clock and space is not an independent container. Space and time affect matter (a ball falls off the table), but matter also has an impact on the structure of space and time. Therefore the latter are themselves *dynamic*. On the other hand, the general theory of relativity remains a classical theory in the sense that it knows no quantum leaps, wave functions, and probabilities.

Now, the belief in the unity of nature is a quasi-religious confession of faith of every theoretical physicist. Can there really be two separate theories in nature? Is it not possible to combine general relativity and quantum theory? To answer this question, let us first take a closer look at Einstein's most brilliant theory.

> Only when the general theory of relativity and quantum theory are based on the same basic assumptions about space and time can there be a theory that unites them.

Einstein's Second Theory of Relativity

Einstein's first theory, the special theory of relativity from 1905, abolished the absolute uniformity and independence of space and time. However, a far more dramatic change in our thinking about space and time was Einstein's second theory of relativity, the *general* theory. It even more profoundly overthrew our popular notions of space, time, matter, and motion, and finally gave up any fixed and static framework of space and time. It constitutes the biggest revolution in our thinking about the cosmos since Copernicus.

Why is general relativity so much more important than special relativity? The structure of space-time in the latter still had a static metric that was independent of the physical bodies moving in it (a metric is a mathematical function used to determine the distance between two points). Although space acts on bodies through effects on their inertia, giving them resistance to any change in their state of motion, the converse is not true: physical bodies do not affect either space or time. The relationship between space-time and physical bodies thus remained asymmetric in special relativity, as it had in Newtonian physics.

> In Einstein's first step, the special theory of relativity, space and time were combined into an interconnected structure of a space-time, but this was still considered to be absolute and unaffected by bodies, motions, or forces.

The general theory of relativity then abandoned the last remnants of the conventional conception of a substantial space and a substantial time. Space-time and all bodies, forces, and motions therein were now integrated into an all-encompassing structure with its own unified dynamics. Masses no longer connect through (gravitational) forces or fields that accelerate physical bodies. Rather, the bodies themselves change the structure of space-time by twisting or bending it. This curvature of space-time in turn affects other bodies and this is how they experience what we call gravity. The classical flat (so-called Euclidean) geometry of space is no longer valid in Einstein's theory. It is replaced by a locally curved geometry whose curvature depends on the mass distribution in the neighbourhood. The physicist John Archibald Wheeler put it in a nutshell:

Spacetime tells matter how to move; matter tells spacetime how to curve.[1]

> Not only does space-time affect physical bodies, but massive bodies also act on space-time. The asymmetry between space-time and bodies in classical physics (and the special theory of relativity) is finally removed in general relativity.

---

[1] K. Ford, J.-A. Wheler, *Geons, Black Holes, and Quantum Foam: A Life in Physics*, New York, (2000), p. 235.

Gravity in general relativity is still caused by the masses themselves. But now they change the geometric structure of a unified, four-dimensional space-time. How can we understand this?

This is often illustrated by an image of a lead ball on a rubber mat. The lead ball causes the rubber mat to deform at the point where it rests. This curvature in turn affects the movement of other balls rolling across the mat. A second lead ball is "attracted" (in the absence of frictional forces) as a result of the indentation created by the first ball. In this thought experiment, the balls on the rubber mat do not actually attract each other through any forces acting upon them, but because they change the shape of the space (which is two-dimensional in the thought experiment).

> What we perceive as gravity on our Earth is in fact caused by the curvature of space-time.

Very Close to the Sun

However, the geometrization of gravitation involves a complication that the example of the rubber mat does not reveal. The initially two-dimensional rubber mat deforms into the third dimension. But the space we consider in describing gravity is already three-dimensional. What should it bend into? We need another, a fourth dimension. And here we reach the limits of our intuition.

The thought that this fourth dimension is time was not new. Einstein had already linked space and time into a coherent four-dimensional space-time continuum in the special theory of relativity, but this space-time did not yet possess its own dynamics. It had to wait for the general theory of relativity to become a dynamic entity. (Of course, the deformation by masses does not take place in the time dimension alone; the four-dimensional space-time as a whole is subject to distortion.[2]) But how is it that we do not notice this dynamic in our everyday life and describe gravity as a force in a static space-time without too much of an error? The answer is that any significant effects of space-time curvature occur only at very high mass densities of a kind that do not exist in our Solar System. However, if we know what to look out for,

---

[2]It does not necessarily take a global fourth dimension to describe the curvature. There are embedded (4 + 1) and non-embedded (3 + 1) theories of general relativity.

we can actually observe the effects of space-time curvature in our close cosmic neighbourhood.

One of these effects that astronomers already knew about before Einstein's time and were unable to explain using Newton's theory of gravitation was the perihelion rotation of Mercury, the innermost planet of our Solar System. Like every planetary orbit, Mercury's is elliptical, so when it circles around the Sun, there exists a point closest to the Sun, called the perihelion. With every revolution of Mercury around the Sun, this perihelion moves a little farther round. However, the observed value of this perihelion precession (rotation) is different from the one obtained by Newtonian physics. For a while astronomers had assumed that this discrepancy must be the gravitational effect of an as yet discovered planet (they already had a name for it: Vulcan). But with Einstein's theory of general relativity, this effect could be explained without the need for a ghost planet.

Many other phenomena in our cosmos which have their origin in the curvature of space-time have been observed over the last hundred years or so since the publication of the general theory of relativity. The most recent such observations concern gravitational waves, whose existence Einstein had already predicted in 1916 and which were observed for the first time in 2015 (Rainer Weiss, Kip Thorne, and Barry Barish were awarded 2017 Nobel Prize in Physics for this). These are distortions of space-time which propagate in a wavelike manner, analogous to the electromagnetic waves of light.

> The existence of a unified space-time and its deformation by massive bodies are experimentally well supported.

Dead End Singularity

Now let us turn to the reason why the quantum field theory of the microcosm and the general theory of relativity cannot be reconciled:

1. Quantum electrodynamics is based on the quantization of the electromagnetic field variables.
2. In order to find a common denominator for quantum electrodynamics and general relativity, the latter would also have to be quantized.
3. General relativity is a field theory in which the field variable is space-time itself.

4. Despite many efforts theoretical physicists have not succeeded in quantizing spacetime.

Einstein's field equations have proven to be so cumbersome and tightly interwoven (they consist of 10 equations for dependent variables that are implicitly related in a complicated nonlinear manner) that the panoply of tricks from quantum field theory have not been sufficient. The technique of renormalization discussed earlier and used to treat the problems of quantum fields in the standard model does not work on the field equations in general relativity. It would take infinitely many modifications of the terms in the Feynman diagrams and therefore an infinite number of indefinite parameters to remove these infinities. And that's just impossible.

> The general theory of relativity is not quantizable. The infinities that inevitably occur upon quantization cannot be spirited away.

Mathematically, the problem arises because the Einstein equations are non-linear. Nonlinearities generally occur, and this is the case here, through feedback loops between the various components of the system under consideration. Physically, this feedback manifest itself in the general theory of relativity in the following way. A mass causes a change in the space-time structure, which in turn has an effect on the mass. The exchange particle in a quantum field theory of gravitation, the hypothetical *graviton*, would thus have to interact with itself (photons do not do this because Maxwell's equations are linear[3]). Upon quantization, this makes the mathematical problem of unwanted infinities extremely unpleasant.

In Einstein's theory of gravitation, the resulting quantum theory is absolutely bristling with infinities. The problems that come with treating charges (electrons) as pointlike can be "calculated away" in quantum electrodynamics. The analogue in a quantized theory of general relativity would be point-like masses. However, their infinities can no longer be glossed over. For given masses, the Einstein equations go crazy below a certain radius (the so-called "event horizon"):

---

[3]Although the gluons and W-bosons in the other quantum field theories of the Standard Model do interact with each other (gluons via their colour charge, W-bosons via their charge), this is not an unsurmountable problem for renormalization. Gluons cannot exist freely because of confinement, and W bosons are very heavy so the weak nuclear force is very short-ranged. The *graviton*, in contrast, would have to be massless like the photon.

- Time disappears (from the point of view of an external observer an object simply stops moving at the event horizon),
- mass density and temperature become infinitely large, and
- the space-time curvature assumes infinite values.

While space-time in existing quantum field theories always remains flat and un-curved (Euclidean), in general relativity its variables collapse into a single point under certain circumstances, so that the individual physical parameters can no longer be distinguished and further calculations become impossible. Physicists speak in this case of a spatiotemporal *singularity*.

> In the general theory of relativity, if a certain radius is undercut for a given mass distribution, or if a certain mass density is exceeded, singularities arise in the space-time structure.

The possibility and the consequences of such singularities within the general theory of relativity were recognized by physicists from the start. It is clear from Einstein's equations that when the entire mass of a body, for example a star, is concentrated in a very small volume, the curvature of space-time and thus the gravitational force becomes so strong that even light can no longer escape. In 1967, the physicist John A. Wheeler gave such a structure the descriptive and suggestive name "black hole", one that quickly made its way into the science fiction literature. However, the radius of such a structure is so small (for the mass of the Earth it would amount to less than one centimetre) or the necessary density of matter so high that it was not initially known how to interpret these solutions. In 1939, Einstein even tried to prove that these singularities could not exist.

> In the end, it is the possibility of black holes that causes the incompatibility of general relativity with any quantum theory.

The Hierarchy Problem

Apart from the obstacle of non-quantizability (or non-renormalizability) of space-time due to the presence of singularities in general relativity, what might a theory that combines quantum field theories, gravitation, and general relativity actually look like? Theoretical physicists are convinced that the path towards such a theory will have to face the *hierarchy problem*, which has already brought despair to several generations of theorists. The background to this problem is that the strengths of the four known fundamental forces in nature show a very clear progression in magnitude. The strong nuclear force is much stronger than the electromagnetic force, which in turn is significantly stronger than the weak nuclear force. But the fourth force, the gravitational force, which so far has been left out of consideration in the Standard Model, is several further orders of magnitude weaker. This hierarchy of strengths leads to some very difficult problems in quantum field theories.

Here is an overview of the three forces so far quantized:

- Strong nuclear force. The basis of its underlying theory, quantum chromodynamics, is the eight-dimensional SU(3) Lie group. Accordingly, there are eight different gluons as exchange particles. They ensure that the quarks stay together and atomic nuclei do not fall apart.
- Electromagnetic force. The basis of quantum electrodynamics is the one-dimensional U(1) Lie group. The corresponding exchange particle is the photon, which attracts or repels charged particles. The electromagnetic force is about 100 times weaker than the strong nuclear force.
- Weak nuclear force. The basis of its associated quantum field theory is the three-dimensional SU(2) Lie group. Accordingly, there are three different exchange particles: W^- particles, W^+ parsticles, and Z^0 particles. These mediate the decay of certain elementary particles. The electromagnetic force is about 10^{13} times weaker than the strong nuclear force.

Physicists would like to describe these three forces in a unified theory, that goes beyond the Standard Model.

> The SU(3), SU(2), and U(1) groups used in the current standard model are not likely to be the last word in a fundamental theory of elementary particles. Physicists are looking for a group that unites *all* three forces.

So far only two of these three forces—the electromagnetic force and the weak nuclear force—have been combined into a single theory, namely the (SU(2)–U(1)) quantum field theory of the electroweak force. This electroweak force splits up into the electromagnetic and the weak nuclear force at certain energy scales. What drives that separation is the mechanism of spontaneous symmetry breaking caused by the Higgs field. This is the same Higgs field that allowed for the possibility of non-zero masses of the elementary particles in the first place (see Chap. 13).

> The Higgs field is the medium that makes the electroweak force appear as two forces, the electromagnetic force and the weak nuclear force, below a certain energy scale.

The next step would be to include the strong nuclear force in this theory. This is not possible against the backdrop of the Higgs field as we understand it today. The reason is that the strong nuclear force is so much stronger than the other two. The symmetry breaking needed to incorporate it (or its separation from a unified force) would therefore have to occur on a much higher energy scale. The enormous energies used to track down the known Higgs field or its manifestations, the Higgs particles, are still orders of magnitude too weak to detect the Higgs field that would be responsible for the spontaneous symmetry breaking of the strong nuclear force.

It is precisely these different energy levels that the fundamental forces operate on that causes the hierarchy problem: a possible quantum field theoretical extension of the Standard Model would require another Higgs field which represents the energy scale of the strong nuclear force and breaks the symmetry of the unified forces on a characteristic energy scale. We shall see in the next section that such large amounts of energy cause other problems than just a high electricity bill.

> The characteristics of all three hitherto quantized forces have not yet been summarized in a single theory. Only on an immensely high energy scale could all these forces be declared manifestations of a single force.

Beyond the Standard Model

Theorists are looking for a better theory, one that can solve all the above problems. Unfortunately, for this purpose, they will be moving into energy ranges that are unlikely ever to be reached experimentally. Of course, this need not prevent theoretical physicists from searching for such a theory. Their hottest candidate for a unifying "Grand Unified Theory", *GUT* for short, is the so-called SU(5) non-Abelian gauge theory, in which the symmetry group is the SU(5) Lie group. Here the three forces acting on the microscopic scale would merge into a single force. However, the SU(5) group reveals its specific characteristics only on a much higher energy scale than the one associated with the Standard Model and today's particle accelerators.

And this is where the real sticking point of the hierarchy problem lies:

- The appropriate Higgs fields are quantum fields.
- Each quantum field generates fluctuations.
- Because the desired Higgs field must be of very high energy, so would be its quantum fluctuations.
- High-energy quantum fluctuations result in high mass contributions to all particle masses, including the known low-energy Higgs particle of the Standard Model (and therefore all other particles).
- With a spin of 0, the low-energy Higgs particle is especially susceptible to such additive contributions to its mass. Nevertheless, its measured mass is only about 125 GeV. So where are the mass contributions from the higher Higgs fields?

The comparatively low mass of Higgs particles can only be explained in an ad hoc manner by the fact that all the contributions of higher Higgs fields cancel each other precisely over a very broad energy range, and that would be too great a coincidence for many physicists to believe.

> The quantum fluctuations of the higher energy fields would have to affect the masses of the known elementary particles. But their measured values are far below the calculated values.

All the difficulties mentioned in finding a unified theory only concern the desire to include the strong nuclear force in a unified theory. In addition, should the gravitational force become part of the new theory, the problems increase even further, because the gravitational force is a further 10^{24} times weaker than the weak nuclear force. The breadth of the energy scale required to unite all four fundamental forces in nature would once again grow by many orders of magnitude. Particle physicists deal with this complication in a very simple way: since gravity plays no role in the micro world, it remains unconsidered.

But there are still plenty of obstacles to a unified theory. The hierarchy problem not only involves the remote energy level at which the much sought-after fundamental force of everything is assumed to break up and yield the three (or four) individual natural forces known today. The respective strengths of the interactions mediated by field quanta, the so-called coupling constants, also differ by several orders of magnitude, and even at very high energies, do not meet at a single point, as they should in a unified theory for a single force. To make things even more complicated, the enormous differences in the measured masses of the many elementary particles are also hard to explain.

> Many things still do not fit together—the energy levels of the symmetry breaking into the three known fundamental forces, the coupling constants, and also the measured masses of the elementary particles resist a uniform description.

Supersymmetries, Strings, and Quantum Loops

A particularly elegant solution to the hierarchy problem could also prove to be a door towards a unification of quantum field theories with gravity. A first step in the direction of unifying all four basic forces could be SUSY, which is short for "supersymmetry," a name that continues the tradition of pompous names chosen by theoretical physicists. While the Dirac theory already doubled the number of particles by postulating an antiparticle for each particle, SUSY goes further and attributes yet another partner particle to every particle and antiparticle known today: each fermion receives a partner boson and each boson receives a partner fermion. The new particles would automatically and precisely cancel all contributions from higher Higgs fields in the Feynman diagrams.

> The supersymmetry theory SUSY predicts a whole series of new particles whose energy scales or masses we do not yet know.

It would be one of the most exciting developments in modern physics if the new particle accelerator LHC at CERN actually discovered SUSY particles. The cheers would be even louder than when the Higgs particle was discovered in July 2012! Most particle physicists believe that SUSY particles do exist. But because no one has any idea what mass they might have, the search looks very much like the famous search for a needle in a haystack.

However, even in a SUSY-extended (SU(5)–) GUT theory, the problem of how to include gravity would remain unsolved. If we are to arrive at the ultimate theory of nature, a "theory of everything" (*TOE*), we must build a bridge from quantum theory to gravity.

> A Grand Unified Theory (GUT), which unites all three atomic forces, would be only a first step on the path towards the Theory of Everythingg (TOE), which would also reconcile gravity.

The most popular version of a TOE today is string theory. It states that the fundamental objects in nature are not zero-dimensional elementary particles without spatial extension, but so-called *strings*, with one-dimensional spatial extent (thereby avoiding the singularities in quantization). The mathematics used in this theory, however, is even much more abstract and more complicated than everything that has hitherto been employed by theoretical physicists for the development of quantum field theories, and it surpasses even the abilities of most PhD students in theoretical physics and mathematics.

Of course, string theory also has a catch: as with all advanced theories of particle physics, including the SU(5) GUT theories, it is not clear that they could ever be placed on an experimental footing. For the energy required to detect such particles will be unattainably high.

> In order to reach the energy scales where physicists' broader theories could show experimentally detectable signatures, particle accelerators the size of the entire universe would have to be built.

Looking Toward the Stars

Because it seems impossible to test the new theories empirically, particle physics is, according to some philosophers of science, in a state of deep crisis. Are string theory and the other extended models of today's theoretical particle physics still science after all? One could say that they are just mathematical metaphysics, as they have long since lost any recognizable relation with the experiential world.

Yet physicists have not given up hope that new particle accelerators will provide them with an unexpected insight into the applicability of their new theories. From the results of the LHC, the biggest and most complex machine ever built by humans, they still hope to gain new insights into physics beyond the Standard Model. Maybe with its help they will indeed find unexpected hints in the direction of one or other theory. So far (first half of 2018), however, it does not look like this will happen.

A new glimmer of hope for particle physicists arises from a completely different discipline, in fact from the area of physics that seems at first glance to be furthest away. In recent years astrophysics and cosmology have made astounding progress. The study of the universe has led to surprising overlaps with insights and questions from particle physics.

> The latest trend in particle physics is to exploit astrophysical findings for research into the subatomic world. This has led to a fruitful cross-fertilization of the two research areas.

The reason for this is as follows. Einstein's general theory of relativity has given rise to an overall cosmological model of the universe. As astrophysicists used their calculations to pursue the history of the universe ever further back toward its very beginning, their equation hit a total singularity, a state in which all the energy and mass of the entire universe were united in a single point. Only the Big Bang could put an end to this singularity.

The remnants of the Big Bang can still be observed today, in the cosmic background radiation, which was detected for the first time in 1964. More precise measurements of this background radiation in the 1990s revealed a fine structure in it with the form of low energy fluctuations. These have their roots in quantum vacuum fluctuations during and shortly after the Big Bang, and are the latest piece of evidence that, in the very early universe, there were regions of energy and matter with different densities, from which galaxies and clusters of galaxies would eventually evolve.

For the question of a Theory of Everythingg the following consideration is of great relevance. In the split second after the Big Bang, the universe was still very small, and yet at the same time it had an extremely high energy, so quantum effects and gravitation must have been working hand in hand. Would it be possible to identify the signature of a common theory here?

Particles and strings on the one hand and the events in black holes and the Big Bang on the other only seemingly describe very different phenomena. Elementary particle physicists and astrophysicists recognize today that the respective limits of their knowledge are the same. It is just that they are looking at them from different sides.

In order to clarify their mutual problems, particle physicists and cosmologists want to bring gravitation and the quantum world together into a unified quantum gravity theory.

Every kid knows about the Big Bang today. However, just the act of naming this event should not obscure the fact that even physicists have no idea how time, space, and matter could suddenly emerge out of this total singularity, i.e., from never, nowhere, and nothing. Only one thing is clear: the developments triggered by the Big Bang ended up 13.8 billion years later by producing a small planet, the third of a small star on the edge of an inconspicuous galaxy, that harboured two-legged creatures with a head that would reflect on the meaning and background of the Big Bang.

Part IV

Cutting Across Philosophical, Aesthetic, and Spiritual Frames of Thought

15

The Path Towards Substancelessness: Breaking with 2,600 Years of Philosophical Thought

For two hundred years, from about 1700 to 1900, classical physics and classical philosophy could hardly be separated. Physics could not manage without a well-defined philosophical basis, and scientists were natural contributors to philosophical debates. The two disciplines had entered into a metaphysical alliance whose credo was:

> Behind the phenomena we observe, experience, measure, and reflect upon stands something independent and unchanging.

Both physicists and philosophers were metaphysical realists. They attributed an independent and absolute existence to things *behind* perceived phenomena.

However, the idea that all being must have a common substantial origin is much older than modern scientific thought. It was the early Greek philosophers who developed this epoch-making principle some 2,600 years ago. They stated that

> ... behind the change of phenomena, just as the very life of nature in summer and winter, blossoming and withering, birth and death (...) is a common ground, absolutely indestructible, eternally immutable ground.[1]

---

[1] W. Capelle, *Die Vorsokratiker. Die Fragmente und Quellenberichte*, Stuttgart 2008.

© Springer Nature Switzerland AG 2018
L. Jaeger, *The Second Quantum Revolution*, https://doi.org/10.1007/978-3-319-98824-5_15

The pre-Socratics described this immutable and absolute reality as *ousia* (οὐσία), a term translated (latinized) into *substance* in Western philosophy. Opinions differed as to what exactly this origin was supposed to be:

- Thales considered water as the constituent that underlies everything.
- Anaximander thought it was the indeterminate, timeless *Apeiron*.
- Anaximenes recognized air as the primordial substance of all being.
- Democritus attributed all being to the action of smallest, indivisible particles, or atoms.
- In Aristotle's thinking, there were four basic substances: earth, water, air, and fire.
- In Plato's philosophy, immutable spiritual ideas representing an objective metaphysical veritableness were the actual reality and common ground of everything.

> The ancient Greeks believed that, beyond our experience there exists a substance independent of all worldly influences. Because this substance is immutable, it is also perfect.

In the course of the history of philosophy, other notions of a substance were added, including the divine principles of medieval scholasticism as an eternal, immutable substance, and since Galileo, the mathematical laws of physics.

In spite of their differences concerning its nature, philosophers agreed that it was only such a substance that could produce observed phenomena, that is, the dependent and unstable perceptions we have of the world. Since ancient times, the Platonic–Aristotelian scheme shaping Western philosophy has thus been characterized by a dualism that draws a sharp line between two opposites, and literally splits the world into two parts:

- On the one hand, there exists a universal and independent *substance*, a kind of primordial foundation that is indestructible and in its innermost essence eternally immutable.
- On the other hand, there is the non-essential, changing, and contingent which philosophers since Aristotle have called *accident* (from the Greek Symbebekos, συμβεβηκός). By this they mean the concrete experiences and perceptions that are subjectively conveyed by our senses, such as colours, seasons, etc., as well as ultimately life and death.

Even today, the idea of substance (and thus the above dualism) continues to determine our everyday perceptions. Consider, for example, a table. If we were to take away all its properties, from its colour to the strength of its material and its shape, we would somehow continue to have the idea that the table itself was still there. One of the most important philosophical discussions of the late Middle Ages, the so-called universals controversy, was exactly about this topic. William of Ockham made himself unpopular when, instead of accepting the general view that the table remains a table even though it loses all its qualities, he put forward an idea of his own: if we take everything away, then nothing remains. For him, the "table" is a term that has an existence solely in our minds.

> In classical philosophy, substance and accident stood in contrast: one was absolute and immutable, the other was what we perceive "only" subjectively.

The philosophical tradition of separating the world into substance and accident was adopted without hesitation by the founding fathers of classical physics, from Galileo to Kepler and Newton. Moreover, the dualism of substantial objects and accidental subjective experiences became the metaphysical foundation of classical physics.

Losing Ground Beneath Our Feet

When physics entered the twentieth century, its practitioners saw no reason to believe that the classical world view of substance and accident would not also guide them on their quest to understand the microcosm. Shortly before, they even thought that they had come very close to discovering the absolute substance, when they discovered the physical atoms. When it turned out that atoms are *not* indestructible and immutable, they just went on searching. Perhaps neutrons, protons, and electrons would turn out to be the long sought immutable substances that make up the world.

But then quantum physics changed everything, bringing with the following novelties:

- wave–particle duality,
- superposition of quantum states,
- spontaneous collapse of the wave function,

- the fact that quantum objects no longer have any independent existence or reality of their own, only potentiality.

How should one judge all these peculiar characteristics of quantum objects? And where was the substance? These were not just questions for physics, but also for philosophy. The discussions about the interpretation of quantum physics in the late 1920s and 1930s, for example, in the Bohr–Einstein debate or attempts to interpret the measurement problem by Erwin Schrödinger, are among the most important philosophical discussions of the 20th century.

And philosophy is still in demand in the 21st century. For example, how can we integrate into our world view the discovery that there are basically no isolated quantum objects? They exist, but not in any independent essence of their own, as their properties only result from their interactions with other quantum objects.

> Permanently fluctuating quantum fields and interactions, which create and annihilate virtual particles, are hard to reconcile with the fundamental ontological assumption that there exists a substance independent of all influences.

Our traditional dualistic thinking is one of the main reasons why non-physicists find it hard to comprehend the concepts and intuitions behind the non-dualistic quantum theory. For a long time, physicists themselves had difficulty understanding it. They simply clung to the usual dualism, albeit in a new form. Bohr's correspondence principle, which separates quantum mechanics and classical physics by the Heisenberg Cut, is nothing but a dualistic concept. And the question of whether an electron is a wave or a particle is a typically dualistic either-or. But physicists were fighting a losing battle. They wanted to illustrate something that clearly cannot be grasped by the dualistic nature of our intuition and thinking.

Philosophically speaking, the elementary particles of modern physics are not substances, i.e., independent things with their own essence, an innermost immutable form, and an autonomous being. Modern physics has even dared to take the next step: it has dropped the very concept of such a substance, and now starts from the premise that reality is determined solely by the *interactions* of particles among themselves. So on this philosophical level, there exist only accidents, no substance. For all physical properties such as mass, charge, spin, and so on are not qualities inherent to the particles themselves, but are defined solely by their interaction with their environment and their functionality. The philosopher Ernst Cassirer formulated this change of perspective in the motto: "*Away from substance, towards functionality.*"

> The transitory interactions between quantum objects have taken the place of an immutable substance as the cause of all phenomena. They are the basis of everything we perceive of the world.

In the most modern philosophical interpretations of quantum field theory, mathematical structures and relations take on the role that physical objects used to have. Such structures are mostly symmetries and invariants in modern physics (see Chap. 18), and they are more reminiscent of Plato's ideas than of any material substance. They play such a central role in the description of the natural world given by theoretical physics today that philosophers speak of an "ontic structural realism". (Ontology is the branch of philosophy that deals with the question of the existence of things in the world. An ontic structural realism is thus the view that existence only arises from structures rather than being a property of things.)

> Structures and relations do not come into the world as a consequence of previously existing things. It is rather the other way around: structures and relations constitute things. Philosophers also speak of "contextuality"

We thus have to say goodbye to the comforting assumption that there is something eternal, solid, and reliable.

Happy Physicists, Unhappy Philosophers

In the first few decades of the twentieth century, physicists had to invest a great deal of intellectual energy to decouple their new concepts and theories from traditional ideas and intuitions, and with the aid of mathematics, find their way into a "substanceless" world view.

The mathematization of worldly objects is not the sole invention of quantum physics, because we already find it in the thinking of Copernicus, Kepler, and Galileo, and even the much earlier ideas of Archimedes and Pythagoras. But there is a big difference between the mathematics of classical physics and quantum physics and the theory of relativity:

- At the beginning of the scientific revolution, the truthfulness of mathematical statements could be more or less directly tested on the basis of perceived reality. For example, the planetary movements could actually be monitored against the stellar background of the night sky. They could be imagined and explained directly and intuitively.
- With quantum physics, and maybe already with the theory of electromagnetism, physicists were no longer able to explain the bizarre phenomena and paradoxes of the microcosm in everyday language. Only by means of highly abstract forms of mathematical description can we grasp the underlying processes of the micro world today, to present our theories and ideas in a consistent way, and in this sense to "understand" nature.

In fact, the abstraction in modern physics looks in every way like a hermeneutic reduction to mathematics ("hermeneutic" refers to the process of understanding, "reduction" means derivation or deduction from something else).

> Only with the aid of very abstract mathematics can the nature of the microcosm and the outcome of experiments be accurately described and predicted.

For the most part, physicists themselves are satisfied with this development, because most of them no longer bother to ask any fundamental philosophical questions. Today's physicists are no longer metaphysicians. And nor do they seek the "true existence and last essence of things," as philosophers have done since the pre-Socratics. On the contrary, their work has led to a concept of substance that seems to dissolve ever further into nothingness.

This literally world-shaking development at the beginning of the 20th century took place far from the public eye. The fathers of quantum physics may consider themselves lucky that their theory was too complicated to attract the interest of the Catholic Church, as other developments had done in the days of Galileo, because the rejection of any independent substance, and thus of a single, eternal truth (see next chapter), represents a much more serious attack on religious dogma than Galileo's new doctrine on the nature of the heavenly bodies.

> The more deeply physicists pushed their understanding of the basic components of matter, the less meaning could be attached to the concept of substance.

But now comes a twist. At first glance, it seems as if physicists had found a way to manage without substances and accidents, by dealing in mathematics. But with the mathematical description of modern physics, the old idea of an immutable substance has sneaked back in through the back door. For most physicists, mathematics is no longer just a tool to understand or describe the world. For them, mathematical forms are, so to speak, perfect ideas—fundamental and valid everywhere and always. Thus mathematics is often considered a substance in the sense of Plato, who had seen the common origin of everything in absolute ideas.

In their vast majority, physicists continue to adhere to a metaphysical belief that there is something fundamental in the things they study and that they are independent of the description we make of them. The only difference is that substance in the form of atoms or particles has been replaced by mathematical concepts such as symmetries, conservation laws, invariants, etc. These mathematical structures are today the bearers of absolute properties (see Chap. 18).

The physicist and philosopher Carl-Friedrich von Weizsäcker expressed the attitude of many of today's theoretical physicists as follows:

> And if you ask, why do mathematical laws apply in nature, then the answer is: because they are their essence, mathematics expresses the very principles of nature.[2]

> Many scientists say that what remains, even if everything else goes away, is the mathematical structure of the laws of nature. They have thus reintroduced a notion of substance through the back door.

"Shut up and Calculate!"

However, mathematics does not offer such a secure foundation as most physicists would like to think. According to its present state, any particle interpretation within a spatiotemporal state description encounters unsurmountable problems. This applies equally to any field interpretation in which spatiotemporally defined classical field values become operators. The notion of a field operator at a specified space-time point cannot be represented in a mathematically consistent way. So it must remain unclear what

---

[2]C. F. von Weizsäcker, *Ein Blick auf Platon – Ideenlehre, Logik und Physik*, Stuttgart 1981.

such operators actually represent physically. As long as there are no practical alternatives, physicists will continue to reckon with their "dirty mathematics" (which includes renormalization techniques in particular). On the other hand, they do this very successfully.

> Quantum physics remains in a state of philosophical chaos. The mathematical particle and corresponding field interpretations of quantum fields are ontologically unsatisfactory.

One question remains: does it even make any sense to ask for a philosophical interpretation of quantum fields? It still seems difficult for physicists not to fall back on a dualistic-substantialist metaphysics. No wonder, for they occupy an uncomfortable spot that falls between two stools: they had to say goodbye to the idea of solid atoms and all material substance, but at the same time a consistent, final theory of everything is not yet in sight. There is not much left to cling to. It may be wise to wait for a definitive unified theory of the microcosm (and perhaps also the macrocosm) that may provide its own ontological interpretation.

16

A New Understanding of Truth: How Quantum Physics Made Absolute Reality Disappear, and with It Absolute Truth

The previous chapter showed that quantum physics was the catalyst for a process in which the traditional philosophical concept of substance would lose its meaning. But it was also the starting point for a further dramatic change in physicists' own concept of physics: the disappearance of any absolute and timeless claim to truth. This chapter explores why the loss of substance should lead to a loss of objective certainty—undermining the traditional notion of a single, ultimate, absolute truth about nature.

Philosophy knows several concepts of truth. One of them describes truth as the agreement of conceptual ideas with the (objectively given) reality. Philosophers also speak of the *correspondence* (or *adequation*) *theory of truth* (*Veritas est adaequatio intellectus et rei*, as Thomas Aquinas put it). This is the prevailing concept of truth in the philosophical tradition, and it is the one that is most seriously undermined by quantum theory.

We have already seen that the pre-Socratics, followed by Plato and Aristotle, created the foundations of a metaphysics that seeks to find the ultimate objective grounds of being and eternally true relationships in nature. This aim is closely related to the ontological–substantial dualism described in the last chapter, which runs like a red thread through Western philosophical thought:

- Only the unconditioned substance that hides behind the phenomena of nature is subject to eternally true laws. These laws allow it to be objectively assessed.

© Springer Nature Switzerland AG 2018
L. Jaeger, *The Second Quantum Revolution*, https://doi.org/10.1007/978-3-319-98824-5_16

- On the other side stands the subjective, conditioned (accidental) human experience, mediated by our senses, which can easily be deceived and therefore must be guided on the path to truth.

> The search for the substance hidden behind phenomena has always been a search for an objective and absolute truth.

The scientific revolution of the 17th century was also deeply permeated by notions of absolute truth. Here, the belief in the perfection of a transcendent (in Plato's case, spiritual) reality, grounded as it was in Ancient Greek thought, had been transferred almost exclusively into religiously determined truths. Early modern scientists, including Newton, Kepler, Galileo, Descartes, and Leibniz, were all devout Christians. For all these founders of modern natural science, the search for truth was closely linked to the glory, perfection, and omnipotence of God and his creation. Their argument was basically that only the Almighty could have created laws of nature that make the world run so perfectly.

> For early physicists with philosophical training, their faith in God served as a deeper metaphysical reason to believe that, beyond our experience, there actually exists a substance which is independent from us as observers.

This belief was not just some personal preference, but actually made modern scientific thinking possible in the first place. Only on the basis of their trust in the divine perfection of an absolute substance could early modern thinkers and their scientific successors remain confident in the search for abstract and universally valid laws in nature. For only with such trust was it possible to believe in the possibility of universal certainty concerning the laws governing natural events. The search for absolute truth gave Kepler the intellectual impetus to develop his theory of planetary motion, but it was also Newton's motivation for his mathematical system of mechanics, and it provided Leibniz with the source and the foundation for his natural philosophy. Without ancient Greek metaphysics and the Christian belief in God's creation, the scientific revolution in the seventeenth century would hardly have been possible.

Some ancient philosophers had already recognized that philosophical difficulties arise with the idea of a substantial and objective nature, i.e., existing independently of us. For example, from the premise that there is an absolute substance, they had proven logically that there must exist smallest, indivisible particles, and at the same time, in an equally logical chain of reasoning, they had shown that these could not in fact exist. But this worried Galileo, Kepler, Newton and their peers little. They never doubted that there was an objective reality and an absolute truth.

> Despite several apparent contradictions, the dream of the unity of the natural sciences, based on the absolute reality of the world and linked by a single common truth, remained a centrepiece or scientific thinking at the beginning of the 20th century.

The Revolutionary from Königsberg

Some philosophers had already shown a good hundred years before the first excursions into quantum physics that the belief in an objective reality, in the existence of things "in themselves", and in a universal and absolute truth could be questioned. One of the first and most important of these philosophers was the German enlightenment philosopher Immanuel Kant. He summed up his thoughts on the topic "What can I know?" in 1781, in his main philosophical work "Critique of Pure Reason". Here he argues that the world and its laws, as we perceive them, are not independent of our experience itself.

The seemingly irrefutable laws of nature are in Kant's view *not* the result of an inherent and objective essence in things that we can somehow perceive the way they are "in themselves". Rather, they are no more than the result of our experiences, which our perception and thought apparatus expose us to, and which in turn shape the objects of our experience.

> Kant says that we never experience the world as it really is; we cannot recognize things as they are in themselves. Their order and structure are not given to us according to absolute, universal rules and laws, but according to our own ways of perceiving and thinking.

The explosive power of this statement quickly becomes apparent: it is not objective and independent things that determine what we perceive. The objects of our perceptions are determined by the apparatus we have at our disposal to obtain knowledge. Kant himself speaks of a *revolution* in philosophical thinking, similar to the Copernican revolution in astronomy.

Because we must turn our ideas about the laws of nature around, i.e., "revolutionize" them (the Latin verb "revolvere" means "to turn around"). Our perception and our thinking are not just passive and receptive. Rather, we ourselves shape the laws governing what we perceive.

A fitting comparison can be made with a bucket into which water is poured. The water must take the hollow shape of the bucket. The bucket thus "thinks": "That must be a law of nature: water *always* takes the shape of a bucket!" Just as the bucket has its given form, our perception and thinking operate in such a way that we can only recognize and describe nature according to the forms of perception and schemes of reason given to us. The astrophysicist Stephen Hawking, who died in 2018, chose the following analogy.[1] He described a goldfish that lives in a spherical aquarium and looks at the world outside through the glass. Inevitably, it sees only a distorted version of reality. At least that is what we think when we look at it from the outside. We attribute to the fish a limited perspective on things. But, strictly speaking, we cannot be sure that our own supposedly undistorted view of reality is somehow "more real" than that of the goldfish. A hypothetical goldfish physicist in his bowl could derive laws of nature about the behaviour of objects outside his bowl that would be just as valid as ours. His world view would be just as real as ours.

> According to Kant, the world does not shape our experience, it is the other way around: our perception, our ideas, and our thinking shape the world we experience.

Among the forms of our perception that shape the world of our experience are, for example, space and time. Kant states Newton's speculation that space and time are absolute and then shows that this cannot be so. He lists the following arguments to support his view:

- Any experience we have necessarily takes place in space and time. Experiences outside of them are not possible. Space and time are required for us to have any experience of nature at all.
- Both are in us before we even have any experience, and both are independent of any concrete experience.

---

[1]S. Hawking, L. Mlodinow, *The Grand Design*, New York (2010).

Space and time are, as Kant puts it, "conditions of the possibility of experience in general". Kant calls such an experience-constituting function "transcendental" (but this should not be confused with the term "transcendent", which describes something otherworldly).

> Kant states that we can never have experiences outside of space and time. Thus space and time are transcendental, i.e., conditions of the possibility of our experience, and we can never assess them in their essence or per se.

Kant does not want to speculate about what happens outside the realm of our experience (this includes, for example, God). But then he nevertheless goes on to talk a great deal about things beyond our limited forms of perception and thought; these explanations even account for the main part of the *Critique of Pure Reason*. He concludes that due to internal contradictions our reason *must* necessarily fail when it attempts to answer questions outside its own limits (which it likes to do when left on its own). Kant calls these contradictions "antinomies of pure reason." They cannot be resolved using pure reason, because beyond our reasoning and our forms of perception, well-known rules such as causality and the spatiotemporal nature of our experiences no longer necessarily apply.

In the brilliance of his philosophy, and deeply rooted in the Enlightenment era as he was, Kant was a radical and revolutionary thinker, and his *Critique of Pure Reason* constitutes a highlight in the history of philosophy. Only in the 20th century did physicists begin to understand the significance of his thinking for their own field.

Happy Philosophers, Unhappy Physicists

After the natural sciences had been pursuing supposedly absolute truths for a long time, the emergence of modern physics initiated a new way of thinking in philosophy. Not least with reference to Kant's philosophy, the idea of absolute reality was systematically pushed back in favour of an empiricist–positivist orientation. The so-called neo-Kantian movement increased its influence within European philosophy until the First World War and also decisively shaped the neo-positivism entering the stage at the same time as quantum physics. One of the most influential circle of philosophers at that time was the "Vienna Circle", which emerged around the physicist and philosopher Moritz Schlick.

The wave–particle duality provides a good example of the manner in which things were moving away from the idea of an objective and absolute reality. At first, physicists still went on searching for an absolute and substantial truth: was an electron essentially a particle or a wave? Niels Bohr ended this either–or situation. In an almost Kantian tradition, his answer was: particles and waves are classical concepts relating to the way we perceive things, and they are no longer valid at the atomic level. So an electron is a particle *and* a wave—and at the same time neither of them, which means that these concepts have no validity, and neither do "both at the same time" and "neither of them" have much significance. Physicists using these terms nonetheless get themselves into conceptual difficulties and indissoluble contradictions.

Unfortunately, this was what happened all too often in the following years and decades. Even in today's physics, there is still talk of waves *or* particles, although any pure particle or wave interpretation leads to insurmountable problems.

> The need to move away from the search for absolute definiteness in physical systems, that is, the search for some kind of true and final characteristics, a need which originated in quantum physics, is one of the greatest philosophical insights of the last century.

Initially, there was opposition to the interpretation that there is no absolute reality in things. The most prominent representative of this resistance was Albert Einstein. In the famous Bohr–Einstein debate on the interpretation of quantum theory, two fundamentally different philosophical concepts confronted each other.

- Einstein called upon the metaphysical foundations of Western philosophy, asserting that the physical world existed independently of the observer. He was convinced that behind all phenomena and measurements lies an objective and independent reality.
- Bohr and his colleagues just wanted to make statements about what can be measured. Everything else they left out. Thus they represented the above-mentioned "empiricist–positivist position". From that viewpoint, any metaphysical consideration of something that transcends the horizon of our empirical experience is considered to be ideological ballast. Thus, it makes no sense to ask what an electron is in itself. One of the most prominent positivists was the philosopher Ernst Cassirer, mentioned in the previous chapter.

> For the positivists, quantum theory is not a theory about an objective reality, but only concerns what knowledge we can have of nature.

For Bohr, quantum theory is in its essence a theory of information, "quantum information". The information and not the object per se is the fundamental quantity to be investigated. For him, quantum mechanical states and the information we obtain about them are one and the same.

It is information that creates reality. And just as in quantum theory physical variables are quantized, information can only exist in multiples of a smallest unit, a quantum, or "bit". In the words of the physicist John Wheeler:

> Every physical quantity, every "it", derives its ultimate significance from bits, binary yes-or-no indications. This is a conclusion that we summarize with the phrase "It from Bit"[2]

Is it therefore so surprising that we encounter *quantum* theories in the smallest structures in the world?

> Quantum mechanics is mathematically complicated and abstract. But in philosophical terms, it came with some significant simplification: what is not measurable, we cannot know about and is thus not considered.

The agreement to rely only on experience or possible physical measurements was a breakthrough. Now the way was free to fundamentally question the classical idea of a strict separation between the independent object and the observing subject. But as soon as physicists left the familiar territory of subject and object, they had to ask themselves: if there is no independent object, then what is reality anyway?

Wine *and* Water

The discussion between Bohr and Einstein summed up the ontological tension that physicists faced in the first third of the 20th century:

---

[2]J. A. Wheeler, "Information, physics, quantum: the search for links", *Proceedings III International Symposium on Foundations of Quantum Mechanics*, Tokyo, 1989, pp. 354–368.

A: Things in the world exist as objects, independent of us human beings as subjects who observe them. This is the worldview of Newton and classical physics, and also the one we use in our everyday lives.

B: An independent objective world does not exist; only our subjective perception (physically, a measurement) gives things their specific state. This is the worldview that the early quantum physicists suddenly had to face.

What's true? A or B? Until about 1930 it was taken for granted that A describes our world. For a long time, physicists tried by every means available to them to maintain the belief in a single reality that exists independently of our observation.

> Einstein's opposition against the Copenhagen interpretation is an example of the conviction that there is an objective reality and therefore an absolute truth.

Even the Copenhagen interpretation had not completely detached itself from this desire for unambiguity, for it considered at least the macroscopic measurement environment as real and objective. In that sense, both Einstein and Bohr were wrong. It was inevitable that quantum physicists had to outgrow the seemingly basic assumptions A *or* B.

The question of whether we prefer answer A or B is not purely academic. It also determines whether or not we claim immutable truths in our worldview (in the sense of an absolute certainty in things and its agreement with our thinking):

- In classical physics only answer A was valid. The world exists as an object, and its properties are fixed, regardless of whether a person is observing it or not. It is therefore clear that there is a single, absolute truth to the functioning of nature (even if we do not always know what it is).
- From the insight of quantum physics that there are no independent, isolated things (particles) and that the separation between subject and object is problematic, a new worldview emerged: A *and* B can be correct—in the macrocosm A applies, in the microcosm B. Truth is ambivalent.

For example, when an electron is measured with the aid of a sophisticated experimental setup, the measuring apparatus can be described using the laws of classical physics. For this, version A applies. For the observed microsystem, on the other hand, version B applies, in which a separation into the independent object and the observing subject is no longer possible. Thus, the concept of reality and truth has fundamentally changed.

There is no objective reality in the micro-world, but only subjective and therefore multiple realities. This does not invalidate the idea of reality per se, but it does undermine the idea of an absolute truth.

In large parts of physics—and in our everyday experience—its classical laws, the classical conception of reality, and the notion of objectivity and determinism remain valid, but not so in others. But how can that be? How does the indeterminacy in the quantum world translate into classical determinacy in the macroscopic world? Most physicists argue pragmatically: even if the separation into subject and object is not really possible, it is still very useful for us in the macro world, maybe even necessary. In our world of everyday experience, the duality of subject and object applies to a very good and adequate approximation. One could also say—following Kant—that in order to be able to perform measurements, the classical laws must apply a priori. (The question of the transition from uncertainty in the microcosm to definiteness in the macrocosm is discussed in Chap. 26.)

But there is also a biological dimension to the question of how and why the world of our experience can be separated into subjects and objects. As the product of evolutionary development, our cognitive functions are adapted to the macroscopic world and not to the atomic or cosmic scales. The evolutionary heritage of our cognitive and thinking apparatus forces upon us the separation between ourselves and external things. We are so to speak programmed to perceive our interior, the subjective, and the external, the objective, as independent in their existence. This subject–object duality is not only useful to us, but vital. The separation between ourselves as subject and our environment as object is an important foundation of our experience, which helps us to find our way around the world. Kant would say that it is a condition for the possibility of experience. The better our ancestors understood the rules in their world on the scales lying between the atomic and interstellar dimensions, along with their objective qualities, the greater were their chances of survival. A human who did not recognize a lion as being real and independent of his or her own perception of it, stood little chance of becoming one of our ancestors.

The strict separation of subject and object is, so to speak, an invention of the macro-world. Our evolutionary heritage forces upon us the separation between the external environment and ourselves. But in the reality of the atomic world such a separation does not exist.

We should ask ourselves why our common, evolutionarily obtained macrocosmic intuitions and conceptions of substance should also be valid on the atomic level, which has always been inaccessible to direct experience.

All or Nothing?

Thanks to quantum physics, a new concept of reality emerged that would rival the traditional metaphysics of Western philosophy. In a worldview that allows ambivalence, *we* decide which viewpoint is "right". Niels Bohr used this idea to arrive at the following statement:

It is the hallmark of any deep truth that its negation is also a deep truth.[3]

However, this idea is not completely new in Western history:

- In the teachings of some early pre-Socratics, the metaphysical duality between subject and object had not yet developed its philosophical dominance.
- As already stated, the concept of reality in quantum physics already resonated in Kant's transcendental philosophy.
- In some respects it finds a correspondence in the phenomenology of Edmund Husserl, which emerged simultaneously with quantum theory. Husserl attributed two dimensions to the process of knowledge: the "act of consciousness" and the phenomenon to which this consciousness is directed.

> The development of concepts of reality or truth made a leap from "A" (there is an absolute reality and thus also an absolute truth) to "A and B" (there are several realities and therefore no absolute truth).

Some physicists and philosophers even go one step further. Not least through consideration of the double-slit experiment, where it depends on the observer whether the electron behaves as a wave or as a particle, they present a completely subjective interpretation of the quantum world.[4] They thus refer to version B alone:

---

[3]As quoted in Max Delbrück, *Mind from Matter: An Essay on Evolutionary Epistemology* (1986), p. 167.
[4]For a modern Interpretation, see C. Caves, C. Fuchs, R. Schack, "Quantum probabilities as Bayesian probabilities", *Phys. Rev. A* 65 (2002) 022305.

- Only an observer allows the observed to exist,
- all experience depends solely on the individual observer, and
- any knowledge can only be justified subjectively.

In other words, quantum objects are solely products of the observation of conscious agents. In this way the subjectivists not only reject the substantial existence of quantum particles, but strip them of *any* reality at all. They are thus true to the dictum: if there is no objective and independent reality, then there is no reality at all.

> Subjectivists throw classical physics, and hence also our everyday experiences, completely overboard. Consciousness possesses a constitutive role for any existence. Truths independent of the subject are excluded at the most fundamental level.

In the subjectivist concept, however, the conscious perception of a measurement plays an all-too-important role. For from the fact that independent, isolated particles do not exist, we cannot necessarily conclude that there exist *no* particles and *no* object *at all!*

The process of measurement is not what generates the quantum objects themselves; it only brings them into a certain state (for example, that of a wave or a particle). Thus, the quantum world does exist, just not with absolute and objectively defined properties.

> The quantum world is real and existent, but without objective and absolute properties. Which of its possible states manifests itself concretely depends on external conditions.

These considerations lead us to interesting parallels with non-occidental traditions of thought in which the subject–object dualism is less firmly rooted in philosophical tradition than in Western thought. Therefore, the next chapter is devoted to possible connections between quantum theory and the philosophy of Buddhism.

17

The Eternal Interplay: Surprising Overlaps Between Quantum Physics and Buddhism

The last two chapters have shown that the fundamental distinction between subject and object in two separate spheres—however obvious it may seem to us—constitutes a philosophical weakness of classical physics. Among other things, it results in a claim about a reality that the findings of quantum physics show us can no longer be unexceptionably supported. But even long before physicists took their first steps into the world of the quantum, representatives of various spiritual traditions of thought objected to the strict division of the world into observing subject and observed object. Ancient philosophers such as Heraclitus, Parmenides, and the Stoics criticized this separation, as did Christian and Islamic mystics (Nicholas of Cusa, Meister Eckhardt, Islamic Sufis), and poets and philosophers of the Romantic era have similarly tried to overcome this strict dichotomy.

Some of the founding fathers of quantum theory knew that the teachings of Buddhism were particularly interesting in this regard. In the millennia-old spiritual thinking traditions of Buddhism, they recognized essential connections with quantum theory.

- Albert Einstein is said to have once confessed:

The religion of the future [...] should be based on a religious sense arising from the experience of all things, natural and spiritual as a meaningful unity.

© Springer Nature Switzerland AG 2018
L. Jaeger, *The Second Quantum Revolution*, https://doi.org/10.1007/978-3-319-98824-5_17

If there is any religion that would cope with modern scientific needs, it would be Buddhism.[1]

- Niels Bohr had a similar affinity for Far Eastern spiritual traditions:

  For a parallel to the lesson of atomic theory […] we must turn to those kinds of epistemological problems with which already thinkers like the Buddha and Lao Tzu have been confronted, when trying to harmonize our position as spectators and actors in the great drama of existence.[2]

- Murray Gell-Mann in his naming for the classification of hadrons was referring to the eightfold path of Buddhism.

In fact, quantum mechanical phenomena such as superposition, non-locality, or the dependence of measurement results on the subject have interesting parallels in Buddhist thinking. Above all, the ideas that caught physicists' attention include:

- abolishing the subject–object duality (there referred to as *non-duality*),
- the non-materiality (conditionality) of all being,
- and the interactions and interdependency of all things and phenomena.

> Several among the fathers of quantum physics recognized the deep connections between Buddhist thought and modern physics.

---

[1]It is not entirely clear whether the quotation in this form is literally from Einstein, or just a reproduction of a similar statement made by him. It is often attributed to the following sources: H. Dukas, B. Hoffman, Albert Einstein: *The Human Side - New Glimpses From His Archives*, Princeton University Press (1954). However, there is no page information.

Here it is quoted by F. Watts, K. Dutton, *Why the Science and Religion Dialogue Matters: Voices from the International Society for Science and Religion*, West Conshohocken 2006, p. 118. It is quite likely that these sentences reflect Einstein's views on Buddhism, since similar statements by him can be found in different places. For example: "Indications of this cosmic religious sense can be found at earlier levels of development - for example, in the Psalms of David and in the Prophets. The cosmic element is much stronger in Buddhism as, in particular, Schopenhauer's magnificent essays have shown us." (New York Times Sunday Magazine, November 9, 1930).

[2]Speech on quantum theory in October 1937 on the occasion of the commemoration of the 200th birthday of Luigi Galvani (Celebrazione del Secondo Centenario della Nascita di Luigi Galvani) in Bologna, Italy. See also: Niels Bohr, *Atomic Physics and Human Knowledge*, edited by John Wiley and Sons, New York 1958), pp. 19/20.

This chapter is dedicated to the following questions: What exactly do Buddhist teachings have to say about the three topics mentioned above? What are the correspondences with and the differences from Western philosophical thought, which still determines our worldview today? And how does Buddhist doctrine relate to the findings of quantum theory?

The End of the Ego Illusion

We consider the thinking of the early Greek natural philosophers 2600 years ago (the pre-Socratics) as the cradle of philosophy. But around the same time, Indian thinkers were asking exactly the same questions and developed as wide a range of answers as their contemporaries around the Mediterranean. Historians assume that—long before Alexander the Great made his way to India—the two cultures strongly influenced each other. Modern analyses of historical sources also clearly show the proximity of pre-Socratic natural philosophy and ancient Indian thought (for example, in the Vedas and the Upanishads) with regard to almost any area of philosophy. The materialistic and sceptical views of those who succeeded them, such as Democritus, Pyrrho, Epicurus, Lucretius, and later Sextus Empiricus, also had strong points of contact with contemporary Indian philosophers.[3]

> The findings of the philosophers of India are clearly related to those of pre-Socratic Greece and later ancient thinkers in the West.

But ultimately their paths would part. As of about 200 BC, despite many commonalities and mutual fertilization, a largely different culture of knowledge and spirituality developed over the centuries in India and China, as compared with the Greek-dominated West and Middle East. A particularly clear example of this is the lack of a sharp separation between substance and accident, mind and matter, consciousness and the outside world, in later Buddhist thought, and putting these conceptual pairs in a nutshell, the lack of any sharp separation between subject and object. Indeed, Buddhist

---

[3]For a more detailed account, the reader is referred to T. McEvilley, *The Shape of Ancient Thought—Comparative Studies in Greek and Indian Philosophy*, New York 2002. This work provides a detailed and insightful comparison between Greco-Roman and Indian philosophy from about 600 BC to 400 AD.

thinking knows no strict dualism of the kind found in traditional Western philosophy, in which all these conceptual pairs are mutually exclusive.

The doctrine of Siddhartha Gautama, whom his disciples called the "Buddha", the awakened, is (like the Greek Stoics) a philosophy of the inner mindset. What is known as insight meditation aims to recognize as an illusion the perceived subject–object separation and ultimately to dissolve it. The Pali word for this practice, *vipassana*, literally means "clarity" or "insight".

> In the teachings of the Buddha, the components of the dualisms perceived by Western philosophy as opposites are actually intrinsically interwoven. Meditation aims at overcoming the distinction between subject and object.

As a consequence, the Buddha also denies the existence of a final and absolute substance. Since there is no separation between subject and object, he refuses to attribute the property of substance *to either of the two*.

Let us first consider the substancelessness of the subject: Buddhism strictly denies the existence of a substantial and irreducible ego. But how does he explain the fact that we humans perceive ourselves as knowing subjects? The Buddha sees an illusion at work here that we should strive to overcome. The ultimate goal of Buddhism is the state of *Nirvana*, in which all factors that bind us to existence have been overcome and extinguished. One of these factors, along with clinging to material things, desire, and greed, is the ego illusion.

> Buddhism rejects the idea of an absolute, independent existence of a substantial ego; with the boundary between subject and object, the ego illusion also dissolves.

The Lack of Substance of the Middle Way

In addition to the existence of an independent ego, that is to say an *inner* substance, Buddhism also discards the existence of any independent and invariable *external* substance: material objects or their underlying components do not independently and absolutely exist. The vast majority of schools of Buddhism teach that there is nothing permanent in the reality we perceive. There is no material substance (or objective substance, in the

Western thinking tradition) and no mental substance (subjective substance) that exists on its own and is independent of anything else.

> Buddhism explains the diversity of all manifestations of material and spiritual nature exclusively through the causal interaction of transient and interdependent entities.

The reasons for this worldview are very simple:

- A substance is absolute and perfect, so it cannot have come from anything else. It should have always been there. But without cause, nothing can be. So there is no substance. (We see that "time" is still used here as a concept in its own right).
- Things can confront us as phenomena only because they do not possess their own, independent being, because they take their form only through their interdependence with other things. If things were independent in themselves, they could not become phenomena through interaction with our perceptive organs.

Western philosophers recognize parallels to the thinking of the pre-Socratics Parmenides and the Zeno paradoxes.

> The belief in an independent and invariable substance is not tenable in Buddhism. Our perception, however, suggests to us a world in which a reality exists independently of our experience.

In the second century AD, Nāgārjuna subjected the principle of impermanence of all phenomena and the lack of substance of all being to a rigorous systematic analysis, Along with the Buddha, he is the most influential thinker of Buddhist philosophy. In his didactic poem *Mūlamādhyamakakārikā* (The Fundamental Wisdom of the Middle Way[4]), he shows how many logical paradoxes arise when one attributes substantial existence to things or their parts. A concrete example of Nāgārjuna's thinking is his reflection on movement in space. He asks:

---

[4]English translation by J. Garfield, *The Fundamental Wisdom of the Middle Way: Nāgārjuna's Mūlamadhyamakakārikā*, Oxford University Press, Oxford (1995).

Is there movement before I lift my foot? Is there a past or future step in which walking would have a beginning?

His example shows that movement can neither be a passage through space that has already been passed through, nor through space that has not yet been passed through. Because there is no room for a third type of space, movement must be impossible. We recognize similarities with the arrow paradox expressed by Zeno of Elea: at any moment on its orbit a flying arrow possesses a specific, clearly defined position. At such a location, however, the arrow is at rest, because in such a fixed place it cannot be in motion. Therefore, the arrow must be at rest at every moment, so it cannot move at all.[5]

Zeno and his teacher Parmenides came to the rather absurd conclusion that all movement is just an illusion. Nāgārjuna chose another way to resolve the paradox. He rejected all independent being and the existence of anything substantial at all. In his teachings of *Mādhyamaka*, the "School of the Middle Path," he explains that phenomena, in their never-ending dependence on conditioning factors, are entirely "empty" (more precisely, "devoid of basic substance of their own"). Nāgārjuna writes:

Not from itself, not from another, not from both, nor without cause: Never in any way is there any existing thing that has arisen.[6]

For Nāgārjuna, there is no independent, autonomous, and immutable substance in worldly being. For him, all things are insubstantial and define themselves only through interactions. He calls this "to be empty".

As an example, Nāgārjuna uses the conceptual pair "walker" and "path walked": without a walker, there is no path walked, and there are no walkers without a path walked. The two can only exist together. In the *Mūlamādhyamakakārikā* there are other examples, such as fire and coal, and seers and sight. Things are always in some interconnected relationship within a fabric of interactions. Nāgārjuna speaks of the principle of the conditional emergence of all being, or *pratītyasamutpāda* in Sanskrit.

---

[5]This paradox can be resolved today using the mathematical concept of limits, which was developed by Newton and Leibniz in the 17th century.

[6]Nāgārjuna, *Mūlamādhyamakakārikā*, 1.

Based on his thesis that everything arises as a function of other things and nothing and nobody possess the property of independent substantiality, Nāgārjuna developed a solid philosophical foundation for the rejection of any subject–object duality:

- No object can exist independently of an observer.
- Subject and object always represent a mutually complementary and inter-dependent pair.

With this view, Nāgārjuna spared Eastern thought a great deal of philosophical hassle.

> Nāgārjuna says that all things are subject to the principle of conditional arising. They have no substance, but undergo a continuous process of change.

The hypothesis of indivisible particles that make up everything also dies along with the rejection of any idea of immutability in things. Nāgārjuna explicitly says that smallest particles are impossible as the primordial substance of all things. Again, the correspondence with Zenon and Parmenides is clear.

Emptiness in the Heart

One of the best-known and holiest texts in Buddhism is the *Heart Sutra*, dating from the 17th century AD, also known as "Sutra on the Essence of Wisdom" (*Prajñāpāramitāhṛdaya* in Sanskrit). The core of this is the teaching of Nāgārjuna about the lack of substance (emptiness) in all being. In its most famous and most quoted sentence, the Sutra says:

Body is nothing more than emptiness, emptiness is nothing more than body. The body is exactly empty, and emptiness is exactly body.[7]

The absence of substance in things is the core of Nāgārjuna's teaching. Unfortunately, the translation of the Sanskrit term *Sūnyatā* into "emptiness"

---

[7]This sentence is also often translated as: "Form is exactly emptiness, emptiness is exactly form, form is nothing more than emptiness, and emptiness is nothing more than form." Instead of "emptiness," some translators also say "lack of substance," which is probably closer to the Sanskrit original *śūnyatā*.

is somewhat misleading.[8] It should actually be translated as "emptiness of any self-existence". For Nāgārjuna does not negate the existence of the body. He rejects only one particular mode of existence, namely an *autonomous* and *independent* existence. This is exactly the meaning of the phrase "the body is empty"—all things are free from inherent and independent being. As Nāgārjuna himself explains, *sūnyatā* does not mean that *there is no being at all* or that things *do not exist at all*. In the body unfolds dependence, hence the absence (emptiness) of absolute properties. This expresses the statement "emptiness is exactly the body".

> The lack of *substantial existence* does not immediately mean lack of existence, and emptiness does not imply non-being, but only non-substantial being.

Nāgārjuna's philosophy is therefore anything but purely subjective or idealistic. He explicitly rejects the view that all perception of external things exists only fictitiously, as a pure illusion or a mere projection in our mind, as some followers of idealistic thought traditions had assumed, including, for example, the English philosopher George Berkeley in the 18th century.

Nāgārjuna tirelessly distinguishes his viewpoint from the extreme position of nihilism, which negates *any* possibility of objective knowledge (and morality), the existence of reality, and even the recognisability of facts. The misperception of Nāgārjuna's doctrine as nihilism is probably a consequence of the Western substantialistic thought tradition, which can but attribute nothingness (Latin: *nihil*) to anything that does not possess any independent existence of its own.

> It is a common misconception of Western interpreters to portray Nāgārjuna's philosophy as nihilistic. He does not deny the existence of the body, only its independent existence.

Nāgārjuna's philosophy is thus located between the following two extreme schools of thought, both of which he vigorously rejects:

---

[8]Nāgārjuna himself said: "Emptiness wrongly grasped is like picking up a poisonous snake by the wrong end." We will thus be bitten!

- In the substance attribute theory, the material entities underlying things possess a fixed immutable existence. They change by external causes only in their respective constitution, but not in their essence.
- In subjectivism or nihilism, there is nothing outside our mind. Recognition and action can only be subjectively justified and validated.

> Nāgārjuna describes both views, the substantialist and the nihilistic, as extremes. For this reason, he calls his philosophy Mādhyamaka, the "Middle Way".

Incidentally, their ideas of empty and not independently existent things led the Indians to an important mathematical concept: the number zero. The Greeks would have found it too strange to give meaning to, or attribute a symbol to, a non-entity, while numbers really do exist. In ancient India, zero meant exactly what it is: the symbol for "no object," a concept that was completely legitimate in Indian philosophy (and mathematics).

Inextricably Linked and yet Not Blurred into One

Nāgārjuna's philosophy must be distinguished from a third school of thought. In a (too) concise and (too) popular colloquial formulation, one could summarize both Nāgārjuna's philosophy and quantum physics with the simple statement: "Everything is connected with everything". But there is a danger here of being understood as making a tribute to some kind of nebulous holism, to which both quantum physics and Buddhism are unfortunately often reduced. The philosophy of Nāgārjuna, although referring to the intrinsic cohesion and inseparable connection of things that are perceived as different, has little to do with any crude form of holism, which only recognizes the existence of systems as a whole, but not their individual building blocks.

> Holism cannot understand systems as a composition of their parts, but only lets them work *en bloc*. Nāgārjuna's philosophy (and quantum physics, too) have little in common with this kind of doctrine of wholeness.

For Nāgārjuna, things are not inseparably woven into a large whole, but neither do they exist separately and independently of each other. Components of a whole do exist, it's just that there are no independent substantial ones.

Take as an example a cup of tea. According to Nāgārjuna, there is no tea in the cup, but only a combination of water, heat, and herbal components (which in turn are each composed of other stuff, that is, they are equally without substance). Even if we generalize the individual parts as tea and thereby relate to a substantial idea of such, neither the concept of tea nor the combination of its constituents has an intrinsic self-nature. Both are in themselves insubstantial.

The same concept applies to interacting quantum systems. As particles in composition with other particles, they do not possess their own independent identity (and therefore cannot be considered as "parts"). Nor do they simply coincide with other quantum objects within a single entity. Their entanglement and indistinguishability from each other determine a connection and dependence that we cannot describe in everyday language.

> Both quantum physics and Nāgārjuna state that all things and phenomena are closely related and contained in each other. Nothing exists apart from anything else, and nothing is completely separated from everything else.

A Bridge Reaching Across More Than 1,800 Years

With the findings of quantum physics, Nāgārjuna's Mādhyamaka philosophy has received increased attention from Western thinkers, because some parallels have become clear[9]:

- The subject–object duality in the measurement process has been rescinded. Upon measurement, there inevitably occurs an interaction and thus entanglement with the measuring system and environment, i.e., the subject and object.
- Quantum objects have no independent existence and no independent substance. In the language of physics, it is not possible to look at a quantum system in isolation.
- In the quantum world, things are fundamentally interdependent. The existence and states of objects and their physical interactions are inextricably linked. It is these interactions that give the objects properties such as solidity, which falsely suggest that there is something substantial behind this solidity.

---

[9]For an excellent discussion, see also: C. Kohl, *Nagarjuna and Quantum Physics: Eastern and Western Modes of Thought*, Saarbrucken (2012).

The correspondences between Nāgārjuna's teachings and quantum physics are astounding, an not only because they show up after a lapse of 1,800 years. Above all, it is surprising that, besides their different cultural and temporal origins, the two approaches have completely different fields of reference and explanation. Quantum physics refers to the microcosm (atoms, electrons, quantum fields, quarks, and so on), while the *Mādhyamaka* philosophy relates to our everyday experience in the macrocosm.

> Cultural and temporal boundaries are superseded, the original aim of the thought system secondary: Nāgārjuna's philosophy can help physicists to understand the nature of the quantum world on a conceptual and philosophical level.

18

Symmetries: Beauty in the House of Physics

From time immemorial, philosophers, spiritual traditions of thought, and even scientists have been asking about the true and ultimate nature of things. Albert Einstein said of the feeling of the mysterious that thereby arises that it was

> ...the most beautiful experience we can have. It is the fundamental emotion that stands at the cradle of true art and true science.[1]

Also Immanuel Kant never lost this feeling:

> Two things fill the mind with ever new and increasing admiration and awe, the more often and steadily we reflect upon them: the starry heavens above me and the moral law within me.[2]

What Kant and Einstein particularly liked was the stringency of the laws of nature expressed through their clarity and beauty. In Kant's case, these were Newton's laws of mechanics, in Einstein's case the equations of his general relativity theory. The language in which this beauty is articulated is mathematics. Galileo put it the following way:

> Philosophy is written in this grand book — I mean the universe — which stands continually open to our gaze [...]. It is written in the language of mathematics,

---

[1]A. Einstein, *The World as I see It*, Citadel Press, New York (2008).
[2]I. Kant, *Critique of Practical Reason*, Conclusion.

© Springer Nature Switzerland AG 2018
L. Jaeger, *The Second Quantum Revolution*, https://doi.org/10.1007/978-3-319-98824-5_18

and its characters are triangles, circles, and other geometric figures, without which it is humanly impossible to understand a single word of it.[3]

> Anyone who has ever felt how elegant and almost wonderfully beautiful a mathematical structure can be, when it captures the essential laws of nature, will never stop gazing in awe.

What indescribable elation Einstein must have felt when he realized that the equations of his general theory of relativity proved to be self-consistent and at the same time accurately described the well-known, but as yet unexplained phenomenon of the advance of the perihelion of Mercury! Such emotions are described by Heisenberg in his autobiography "Physics and Beyond":

> At the first moment I was deeply shocked. I had the feeling to look through the surface of atomic phenomena down to a ground of strange inner beauty lying deep beneath it and I nearly got dizzy from thinking that I should now go into the matter of investigating this abundance of mathematical structures nature had spread down there in front of me.[4]

This sense of beauty that scientists experience in particular moments like these, when reading the message of elegant equations, resembles the feeling that overcomes artists when they create something that especially satisfies their aesthetic sense. In fact, there is a common denominator: symmetry.

> Theoretical physics of the twentieth century discovered symmetry as a central principle that could guide them in their quest for knowledge and this gave them a fundamental belief in the unity of nature.

From Art to Science

The term symmetry derives from the ancient Greek *symmetría*, a combination of *syn* (together) and *métron* (the right measure). Symmetry means equal or even measure. It is akin to the Greek word *harmonía* (same proportion).

---

[3]Galileo Galilei, *Il Saggiatore*, Rom 1623, cit. after R. Popkin, *The Philosophy of the Sixteenth and Seventeenth Centuries*, Simon & Schuster, New York (1966).

[4]W. Heisenberg, *Physics and Beyond: Encounters and Conversations*, New York (1971), p. 78.

In the ancient, medieval, and early modern conception of the arts, symmetry described the ideal proportions of length and distance in sculptures, paintings, or buildings. The human body was considered a good model for harmonious numbers. For example, the ratio of the length of the arm to the entire body is almost exactly one quarter. And with outstretched arms and legs, their ends can be inscribed exactly in a square and also a circle, with the navel at their centre. The corresponding drawing by Leonardo da Vinci is well known, and it made him one of the most famous artists and scientists of the Renaissance.

In addition to ideal proportions, ancient thought was familiar with two other concepts of symmetry: the mirror-image symmetry as it finds its expression in the left and right halves of the body, and the balance of opposites, as articulated in Greek medicine and its theory of body fluids.

> Most ancient and modern art conceptions recognize symmetry as an essential criterion for beauty and perfection.

Modern physics is also guided by symmetry considerations. Its goal is to work out the processes and structures underlying the confusing complexity of natural phenomena. Yet most physicists cherish the deep belief that nature, despite the diversity of its phenomena, can be proven to be simple on a fundamental level. In this simplicity, as it finds its expression in mathematical structures, nature shows its true beauty.

What nature directly offers us and our senses, however, is anything but simple. Rather, scientists must first separate the colourful and confusing mixture of phenomena, freeing the important features from any unnecessary adjuncts (such as friction during free fall), until the simple underlying processes are revealed (e.g., the law of free fall). Only that simple part then appears to us as beautiful.

Historically, this focus on the essential has been particularly successful in the field of astronomy, because in space there are hardly any disturbing effects like friction. It is largely for this reason that astronomy represents the starting point of the scientific revolution. Johannes Kepler was so enthusiastic about the beauty and simplicity of the planetary motions that he claimed to have recognized the highest divine principles in the laws he discovered.

> Symmetry is beauty—and this is also true in physics. For a physicist, the beauty of the laws of nature shows up in their graceful simplicity.

Although the symmetry concepts of art and physics overlap, they have different emphases. In physics, it is less about proportions and equilibriums than about order and structure. Werner Heisenberg explicitly elaborates on the demand for symmetries in modern physics:

> The final theory of matter will, like Plato's, be characterized by a series of important symmetry requirements.[5]

But these symmetries are no longer necessarily intuitive, as he goes on to say:

> These symmetries can no longer be explained simply by figures and pictures, as it was possible with Platonic bodies, but by equations.

In science, symmetry stands for a fundamental structural order.

Symmetry as Invariance

Science is beautiful when trivialities are cut out and every piece of the mosaic of knowledge takes its correct place in the overall composition. The laws are clear and understandable, they express a harmony, in which the fundamental order and structure of nature becomes visible. This beauty is perceived as symmetry. However, the concept of symmetry in science is a bit more abstract than mirror images or point symmetries of the kind we know from school. Its origins date back to the 18th century, when scientists started classifying crystals.

The symmetries of crystals are manifest upon rotations about certain axes and angles which do not change their appearance. Thus, as already noted by Kepler, snowflakes, for all their individuality, always display the symmetry of a hexagon; they can be rotated by 60 degrees without changing their appearance. This can be explained today, although Kepler did not know this, by special properties of the water molecule. For the cubic crystals of common cooking salt (NaCl), it is 90° rotations that preserve the crystal's appearance. In the 19th century, these very concrete symmetries led to a generalized definition of symmetry: invariance under transformations.

---

[5] W. Heisenberg, *Steps Across Borders*, Munich (1971).

Just as crystals rotate and reflect in certain ways, so do other things in nature. For example, the two wings of a butterfly are mirror images of each other. This means that the reflection of one wing on the central axis of the butterfly yields precisely the other wing. Physicists say that the image of the butterfly is invariant under the mirror image transformation.

> A body is considered symmetric when certain transformations, such as rotations or reflections, transform it into the same shape.

So far we have distinguished three forms of symmetry:

* Harmony: a concrete symmetry like correct proportions (in art)
* Reducibility: an abstract symmetry like simplicity (in science)
* Invariance: the concrete symmetry of crystals and other bodies that do not change their shape under certain rotations or reflections (cube, hexagon, etc.)

And now comes the leap into the abstract depths of physics and mathematics. It is not only bodies that can be characterized by those transformations that leave them invariant: the same applies to mathematical equations. They, too, can retain their form under certain transformations. Nineteenth-century mathematicians, led by the Frenchman Évariste Galois (for algebraic equations) and the Norwegian Sophus Lie (for differential equations and general geometric structures), developed an entirely new mathematical discipline that followed the scheme of crystallography. It became known as group theory, although this name does not have much in common with the everyday use of the word "group". Rather, a mathematical group is a set—here, a set of transformations—with a rule of combination satisfying certain properties.

The symmetry group of an object, whether it be a geometric entity or an algebraic or differential equation, consists of those transformations that leave the object invariant, along with a specific rule that defines the combination or *product* of two transformations. For example, all rotations of a regular plane polygon with n sides which map the figure back onto itself form a group (in this case, it encompasses all rotations through multiples of the angle $360/n$ degrees). The group product is the application of two consecutive rotations, which results in a single rotation by the sum of the two angles, also an element of the group.

The transfer of the concept of symmetry from concrete geometrical objects to abstract algebraic structures made symmetry a core principle of theoretical physics.

For most bodies, their symmetry group is determined by individual, i.e., discrete angles, through which the body can be rotated without changing. However, two figures remain invariant when rotated through *arbitrary* angles: the circle and the sphere. Groups in which the angles or other quantities characterizing it are described by *continuous* parameters are something special in physics. They are called *Lie groups*. The reader will remember them from Chap. 13, where such a group characterized Gell-Mann's classification of hadrons. The group describing the symmetry of the sphere in three-dimensional space is a Lie group called SO(3). For its part, it is three-dimensional, too. In Chap. 13 we also mentioned another Lie group, the eight-dimensional SU(3) group. It does not refer to *real* rotations, i.e., rotations in our everyday three-dimensional space of real numbers, but to rotations in a *complex* three-dimensional space, i.e., a space based on complex numbers. As a reminder, complex numbers are numbers that actually should not exist according to our intuition—they include numbers whose square has a negative sign—but they can be described consistently in mathematical terms.

Although abstract spaces and their Lie groups may seem like metaphysical sorcery to non-mathematicians, they play a very concrete and important role in today's theoretical physics.

The Greatest Female Mathematician of All Times

In short, the vast majority of symmetries in physics are described by Lie groups. They include transformations under which the fundamental physical equations remain unchanged. What is special about these Lie group symmetries is that they are related in a rather surprising way to so-called *conserved quantities*, i.e., physical variables that do not change their value when a system undergoes any process whatever. This relationship, which may sound unspectacular, has a meaning for physics that can barely be overestimated. Here are a couple of examples:

- A simple shift of the time variable has no influence on the laws of nature. An apple will fall off the tree today according to the same rules as it would do so tomorrow or yesterday. What sounds simple and self-evident has a great significance for our world: the fact that the laws of nature do not change over time implies the law of *energy* conservation: in a physical system the total energy never increases or decreases, but always stays the same.

- The same applies to shifts in space. If, instead of the variable x, one chooses the variable $x + \Delta x$ (where Δx is a spatial displacement in any direction) and considers the resulting physical equation, this equation retains the same form. That must be so, otherwise different physical laws would apply in New York and Paris. The fact that the equations of physics do not change under spatial shifts implies the conservation of the *total momentum* of a system.

> In addition to energy and momentum, for example, angular momentum and charge also belong to the set conserved physical quantities. Conserved quantities are, so to speak, the rock in the surf of world events.

And now comes the surprise: *each* physically conserved quantity can be assigned a symmetry which can be described mathematically by a Lie group. For physicists, this comes with a great advantage. For from the properties of the Lie groups, they can derive the type and properties of the conserved quantities. And because the opposite applies equally, every Lie group symmetry leads to a conserved quantity, and physicists can specifically search for conserved quantities by searching for new Lie group symmetries.

The macroscopic conserved quantities such as energy, momentum, angular momentum, etc., are (most likely) all known. Each of them can be assigned a special Lie group under which the equations of the corresponding physical theory remain invariant. In particle physics, on the other hand, there might exist some symmetry or other and hence a conserved physical variable that is not yet known. Lie groups have thus become important tools for physicists in their search for new elementary particles and their properties.

> If a theoretical physicist discovers a new Lie group that expresses a symmetry (invariance), this indicates a correspondence in the structure of nature. Its characteristics will serve as a guide to the properties of the conserved quantity to be discovered.

It was a woman who first recognized the relationship between conserved quantities and the symmetries in the mathematical equations of physics. In 1918, the German mathematician Emmy Noether formulated the theorem named after her today, which links physically conserved quantities with the invariances of the basic physical equations under defined transformations. For physicists, the Noether theorem is like a lighthouse that guides them through rough waters. With the help of its predictions, they have celebrated many discoveries over the years.

- The eight different gluons of the strong nuclear force are a direct consequence of the SU(3) group being eight-dimensional.
- Analogously, the three exchange particles of the weak nuclear force result from the three-dimensional structure of the SU(2) Lie group.
- From the properties of the SU(2) group, it can also be deduced that an integer spin always remains integer, and a half-integer spin always remains half-integer. This explains the existence of the two fundamentally different types of quantum particles: individual fermions always remain fermions and individual bosons always remain bosons (although fermions can couple to form bosons, as for example in superconductivity).
- The properties of the photon correspond to the U(1) Lie group.
- The fact that gluons and vector bosons interact with each other, whereas the photon does not, can be explained by the specific properties of their underlying Lie group symmetries.

Plato would have been thrilled if he had known about this wonderful connection between mathematical symmetries and concrete manifestations in the physical world.

> Through the Noether theorem, the completely abstruse, highly complex, and abstract concept of Lie groups has acquired a very concrete relevance in the physical world and has led to many significant discoveries in physics.

An Excursion into Fourth Year University Mathematics

Now comes a little additional information for readers who want to venture further into the abstract terrain of group theory. What precisely are the various symmetries the Noether theorem refers to? And what makes them Lie groups?

Reflections, spatio-temporal shifts, and rotations still appear quite intuitive to us. The symmetry groups in quantum field theories, on the other hand, can consist of more abstract symmetries. For example, a particle with spin ½ does not return to its original state after a rotation of 360 degrees, but only upon a rotation through twice the angle, i.e., 720 degrees (this effect can be illustrated with a Möbius strip, around which one has to walk twice in order to come back to the starting point).

Physicists also speak of "gauge symmetries", a name introduced by the German mathematician Hermann Weyl (the same one we encountered earlier as the lover of Schrödinger's wife) in the 1920s. This term means nothing other than that certain quantities in a theory can be freely chosen, physicists say "gauged", without changing the basic physical laws.

> The symmetries taking the form of Lie groups in quantum field theory (and some even in their classical predecessors) are called gauge symmetries.

The concept of gauging is well known in measurement technology. For example, a weighing scale must be calibrated (gauged) with the help of a test weight and a thermometer by a reference temperature. What exactly is a kilo anyway? And what does one degree Celsius mean? Basically, their definitions are completely arbitrary, as evidenced by the different temperature scales—Celsius, Fahrenheit, and Reaumur. The same is true with the classical equations for the electromagnetic field. Here, a certain field function can (and for concrete calculations, must) be freely selected. Here, too, physicists speak of a gauge.

Here is an example from quantum physics. The wave function can be multiplied by a (complex) factor with certain characteristics (its absolute value taking the value of one) without changing the form of the Schrödinger or Dirac equations. Physicists also speak of an invariance under a (local) phase transformation (i.e., the concrete factor may vary from one space-time point to another). The set of all these transformations forms a specific group, the U(1) Lie group.

- The conserved quantity corresponding to this U(1) Lie symmetry group is the electric charge.

Other gauge symmetries relate to more complex "charges" that also constitute conserved quantities:

- The gauge symmetry of the strong interaction in the standard model is the SU(3) Lie group. Here it is the "colour charge" of the quarks that is conserved.
- By analogy, the symmetry group of the weak force is the SU(2) Lie group. The corresponding conserved quantity is called weak isospin.

Each conserved quantity in quantum field theories corresponds to a gauge symmetry, and vice versa. The symmetry groups give the corresponding quantum field theory in the Standard Model its special features.

Too Good to Be True

Now back to everyday business. If we wish to impute a metaphysical faith to physicists, it must surely be their deep confidence in the symmetries of the laws of nature. Perhaps the person who most radically articulated this belief was Paul Dirac:

It is more important to have beauty in one's equations than to have them fit experiment.[6]

And for the French mathematician and physicist Henri Poincaré:

If nature were not beautiful it would not be worth knowing, and life would not be worth living.[7]

Up to this day, the unification of quantum mechanics and special relativity derived by Dirac on the basis of purely theoretical considerations of symmetry (with regard to the symmetries of the Lorentz transformations) is regarded as one of the most impressive examples of mathematical elegance and beauty in physics. From it there followed predictions that were as astonishing as they were impressive, one of them being the existence of antimatter, and the Higgs particle was already postulated on the basis of symmetry considerations in the 1960s. Physicists were so sure of their theory and thus of the existence of this crucial particle that they were ready to wait half a century for experimental proof of its existence—and to persuade the relevant political decision-making bodies to spend the many billions of dollars required for this task.

However, this search for symmetry has a downside. Precisely *because* physicists more or less construe symmetries as a metaphysical "principle of true being," there is a danger. The argument that something "is true because it is beautiful" has similarities with medieval scholasticism. Back then, the argument went as follows:

---

[6]P. Dirac, *The Evolution of the Physicist's Picture of Nature*, Scientific American, 208 (5) (1963).
[7]N. Poincare, *Science et méthode*, Paris (1908), p. 19.

- God is true because he is the principle of all being,
- and he is the principle of being, because he is true.

Such a circular conclusion is of course not a proof of God's existence.

> The symmetry requirement in physics can easily lead to circular reasoning. It cannot be the sole quality criterion of a scientific theory or even ideal of science.

In today's discussion of supersymmetry (SUSY) and supersymmetric quantum field theories, some theoretical physicists seem to have been beguiled by a very similar error of thought. For, despite great effort and expense, we have yet not found the slightest experimental evidence for SUSY particles. They then try to reconcile these negative results with their theory using more and more ad hoc explanations.

This is reminiscent of attempts in the Middle Ages to uphold the Ptolemaic worldview. Astronomers came up with ever more elaborate theoretical complications to support the notion that the Earth is the centre of the Universe, despite an increasing number of observations contradicting this model. In contrast to the medieval scholars, what prevents physicists from drifting too deeply into metaphysical speculation is the requirement that their theories obtain experimental validation, as required by the scientific method.

> "Beautiful" theories should not be a priori immune to criticism. Yet, there is a tendency in modern theoretical physics to cling too much to "simple" and symmetrical theories, even if experiments seem to refute them.

The Birthmark on the Cheek

Let us now leave the abstract symmetry in mathematics and return to the concrete symmetry of physical bodies. Impeccably symmetrical forms are omnipresent in art and architecture. But artists and art historians largely agree that perfect symmetry has a certain sterility to it that can run counter to our aesthetic sense. In pictures and sculptures, highly regular geometric bodies may seem rather uninteresting. Thus writes Kant in his "Critique of Judgment":

All stiff regularity (such as that which borders on mathematical regularity) is inherently repugnant to taste, in that the contemplation of it affords us no lasting entertainment [...] we get heartily tired of it.[8]

Even completely symmetrical faces are hardly considered attractive; they rather irritate us. You can try this out by mirroring half a face of a portrait to create a whole face. The essential and interesting thing about symmetry is rather that we perceive any breach in it particularly strongly. A small break in symmetry, be it a slightly crooked smile or a raised eyebrow, makes a face much more attractive. Likewise, works of art usually only receive their uniqueness through deliberate breaking of symmetry and order. Too much asymmetry, likewise, appears chaotic.

This too is a facet of symmetry: if it is too perfect, it disturbs our aesthetic sense. Only from a balanced break with the static and familiar symmetry can something new and interesting unfold.

And even in physics there is no absolute symmetry. Thus, at the very heart of our current most fundamental theory, the Standard Model of elementary particle physics, we encounter a striking breaking of symmetry. If the (gauge) symmetries of modern quantum field theories were completely intact, i.e., if their equations were completely invariant under the corresponding (gauge) transformations, there would be no mass in the world. The masses of the elementary particles arise only from the violation of symmetries by the mechanism of "spontaneous symmetry breaking", also referred to as the "Higgs mechanism". This was named after Peter Higgs who, with some of his colleagues, introduced it into the mathematical–theoretical description of the elementary particle world in the 1960s (for which he was awarded the Nobel Prize in Physics in 2015, after the Higgs particle had been detected).[9]

Of course, that does not mean that symmetries lose their relevance in physics. The process of spontaneous symmetry breaking does not make all of nature asymmetric. On the contrary, it is the mathematical symmetries in

---

[8]I. Kant, *Critique of Judgment*, Translated by J. C. Meredith, Oxford University Press, New York (1952), p. 73.

[9]The phenomenon of spontaneous symmetry breaking also occurs in other contexts, e.g., in ferromagnets, in the formation of crystals from a liquid, and also in superconductivity. There it is described by the Landau theory, named after the Russian physicist Lew Landau who developed it in 1937.

the field equations of physics that make symmetry breaking possible in the first place. In fact, the antagonistic pair "symmetry–anti-symmetry" yields a completely unique dialectic in art as well as in science: symmetry breaking requires a basic symmetrical pattern that can be broken in the first place. Only from a well-known and easier-to-grasp symmetrical basis can whatever is individual and complex in nature unfold in its unique way as a deviation from the symmetrical norm.

In physics as in art, the little irregularities are of great importance. It is only from a breaking of symmetry that certain elementary properties of nature can arise, including the mass of all matter.

19

Quantum Consciousness and the Tao of Physics: On Quantum Holism, Quantum Healing, and Other Quantum Nonsense

Despite its exciting philosophical and technological implications, quantum theory becomes understandable only in a language that most people have little liking for and even less mastery of: mathematics. Unfortunately, without a few years studying mathematics, no one is likely to get too far into it (or rather, too deeply into it). As a result, the theory that threw classical physics off the throne and shattered a whole metaphysical framework, thus relativizing 2,600 years of philosophy, is hard to understand for non-physicists at its most profound levels. And unfortunately again, this can open the door to misunderstandings, misinterpretations, and sometimes misuses of quantum theory.

A highly diverse group of people refer to quantum physics without having fully understood it; including, above all, those who describe themselves as "spiritually oriented".

Especially among those with spiritual interests, there are many who are looking for a scientific foundation for their none too scientific worldview. For example, there are those that hope that quantum physics will provide the desperately desired solution to the so-called mind–body problem.

Nobody knows how matter (our body) and mind (our soul or consciousness) are related. This question about the nature of our mind is one of the fundamental philosophical questions that has kept mankind busy for millennia. How can it be that a bunch of molecules in our brain possess consciousness? It seems clear that our thinking and experiencing cannot be explained solely by

© Springer Nature Switzerland AG 2018
L. Jaeger, *The Second Quantum Revolution*, https://doi.org/10.1007/978-3-319-98824-5_19

what happens in a few chemical reactions, that is, by purely material processes. For even with the most sophisticated measurement methods, we have so far been unable to reduce subjective moments of our experience entirely to objective (measurable) conditions (for example, brain states).[1]

Many philosophers, therefore, support a notion which they call "the irreducibility of the subject." They say that, if no connection can be found between mind and body, the spiritual must exist independently of all that is physical. And some go even further, claiming that the search for an explanation of the way matter manages to "think" could turn out to be completely the wrong approach. Who says it is not the other way around? Not matter that determines the mind, but mental or spiritual entities that create the physical world.

Immanuel Kant may even be the considered as the instigator of this argument. As the reader knows from Chap. 16, Kant claimed that the things we perceive are already shaped by our own intuition and the categories of our thinking. This is the reason why nature can be described according to mental principles, such as mathematical laws (mathematics is a product of our mind, so goes the argument). And then comes the daring conclusion by the spiritually minded (where any reference to Kant gets lost): the spiritual determines the physical, the material is in fact built up of elementary spiritual entities.

> One approach to philosophy claims that it is not matter that determines the mind, but rather the mind that determines matter.

One may find reasons to share this worldview, and others to reject them. But now comes the sticking point: these are philosophical games of thought, and quantum physics has nothing to do with all this. And yet some people try to use quantum physics as a basis for putting out their particular views in this game. They believe they have finally found the missing link between body and soul, and at the same time the proof that the immaterial dominates the material.

Quantum Physics and the Mind—A Popular Connection

Those considering that spiritual principles are the foundation of the material world and that quantum physics constitutes the hinge between the two are able to refer to some prominent names. The German psychologist and

---

[1]The measurement of brain states, however, already enables us to determine the content of thoughts and the emotional focus of a given person.

theologian Frido Mann, grandson of Thomas Mann, and his wife Christine Mann, also a psychologist and daughter of Werner Heisenberg, recently wrote an entire book about it (in German): "Es werde Licht" ("Let there be light").[2]

They argue, in agreement with the general consensus, that our world can be grasped and described with the help of intellectual principles, such as formulas and laws. This is precisely what physicists and mathematicians actually do. But the Manns go much further, and say that the material world is a *direct imprint* of the mental. Only the mental, our thoughts and perceptions, make the world real.

> "I think therefore I am," Descartes once said. However, some adepts of pseudo-quantum philosophy claim, "I think, therefore the world is."

"Our thinking changes reality," say the Manns. The idea that our thinking changes our *perception* of reality or our perception of real life is indeed common experience. But that is not what the Manns are referring to. They are convinced that our thinking has a *direct impact* on the physical–material world. This would mean that the course of an iron ball on a sloped surface changes if we just think about it or look at it. Common sense immediately rejects that statement: how could that be? The answer given by the Manns and their colleagues is that mental entities (and thus also our consciousness) and matter consist of the same basic substance, so-called *quantum information*. Put simply, if someone is able to hit a nail with a hammer, they can equally well do this with their thoughts.

Quantum Physics as a Substitute for Clear Thinking

The idea that quantum *information* is the foundation of our world has its origins in physicists' understanding that, beyond our measurement, a quantum object has no independent existence. Only when there is information about the position or state of an electron does it become "visible". So when physicists talk about quantum information as the basis of matter, they mean the quantum physical worldview of a material nature constructed out of quanta, objects such as electrons and other atomic particles.

---

[2]F. Mann, Ch. Mann, *Es werde Licht*, S. Fischer Verlag (2017).

Quantum mechanical states and the information we have about these are one and the same. By "information about the electron", we mean nothing other than the electron itself.

The physicist Thomas Görnitz, to whom the Manns refer, comes to the conclusion, however, that abstract and free-of-meaning bits of quantum information (so-called "AQI bits") can be recognized as the foundation of cosmic evolution. These AQIs establish a quantum pre-structure Görnitz termed "protyposis" (Greek for "pre-formation"). For him, this simplest quantum structure forms the basis for a unifying scientific description of both matter and consciousness. Matter turns into—as Wörner calls it—"formed" quantum information.[3] Sounds strange? Well, it is indeed! Not many physicists and philosophers would support Görnitz's proposal.

> For some people, the world is not made up of (material) atoms, but of an (immaterial) mental substance called "quantum information". But only very few physicists support this idea.

A Few Decades Too Late

The notion that the world we experience does not exist as we perceive it, but that our mind creates it in this form—in the modern version of this idea with some quantum information—has been around in philosophical and literary circles for a long time, and has always found an (albeit limited) number of followers. For example, George Berkeley expressed the view that all our perceptions of an outside world are brought into being only through our consciousness, and that there is no material world beyond a perceiving mind. Such worldviews were enthusiastically received by movements like the German idealists and the poets and romantic thinkers of the early 19th century. Even in many of today's spiritual circles, followers of this view are strongly represented, and they feel supported by the early interpretations of quantum physics.

---

[3]See also (in German) Th. Görnitz, B. Görnitz, *Von der Quantenphysik zum Bewusstsein – Kosmos, Geist und Materie*, Springer-Verlag (2016). English papers on the topic are: Th. Görnitz, *Quantum Theory as Universal Theory of Structures—Essentially from Cosmos to Consciousness*; to be found at https://cdn.intechopen.com/pdfs-wm/28316.pdf; and Th. Görnitz, *Simplest Quantum Structures and the Foundation of Interaction*, Reviews in Theoretical Science, Vol. 2, Nio 4 (2014), pp. 289.

> The idea appears obvious: if an electron has no properties independent of observation, then that also applies to a tree, a table, etc. Without observer, none of them exist, right?

Part of the guilt for this philosophical mess can be attributed to certain aspects of the most popular philosophical interpretation of quantum physics to date: the Copenhagen interpretation, which the reader is already familiar with. On the basis of:

1. the impossibility of assigning a reality to quantum objects in the immediate sense, and also
2. Heisenberg's uncertainty principle,

their representatives, grouped around Niels Bohr and Werner Heisenberg, felt compelled to conclude that physical variables and measured quantities in the microcosm acquire their properties and values, indeed their very existence, only by being measured. Thus, these pioneers of quantum physics raised the question of whether it is not in the end our conscious perception and our will which make the wave function collapse during measurement, and are thus responsible for the objectivity of things. The reader will encounter this idea again in Chap. 22, in the discussion of "Wigner's friend." Christine Mann (born Heisenberg) considers her own father, the founder of the uncertainty principle, among the pioneers who thought this way. Whether she is right or not is controversial among historians of science.

> Indeed, the pioneers of quantum physics were asking whether human consciousness plays a key role in the physical processes of the micro-world, or at least our experience of them.

What the followers of the idea that mind creates matte do not consider is that physics has made some significant progress in the meantime. Through the interplay of abstract mathematics and highly skilled experiments, physicists have meanwhile been able to gain a much more precise idea of the quantum-mechanical measuring process (see also the entire fifth part of the book, especially Chap. 26).

Richard Feynman said in 1967:

> There was a time when the newspapers said that only twelve men understood the theory of relativity. I do not believe there ever was such a time. [...] On the other hand, I think I can safely say that nobody understands quantum mechanics.

Well, *today* there are quite a few who can claim to understand quantum physics (and Feynman was certainly one of them). Most contemporary physicists refrain from stating that the mind or human consciousness plays a role in quantum physics, or that quantum properties play a role in consciousness. With the exception of a thought experiment by Eugene Wigner (see Chap. 22), the mind or our consciousness simply does not appear in physicists' interpretation of quantum theory. The solution to the mind–body problem most likely has to be sought elsewhere.

> Anyone who still assumes that the (macro-) world does not exist without observers and calls quantum physics to the witness stand to support this belief, has failed to take note of the advances in physics in recent decades.

A Child's Play

Children sometimes play the following game. One child, the ox, keeps his eyes shut, and the others are then allowed to move. As soon the ox open his eyes, the other kids must stop. Whoever the ox sees moving is out. In a world in which we only create reality through our observation, it would be the other way round: only when we look, does something happen.

What speaks against such a worldview in *everyday life* is that we definitely do not experience things changing their course just because they are being observed. An iron ball is an iron ball with all its characteristics, whether you watch or think about it or not. "Well, you look at it when you say that, don't you?", one could counter. Strictly speaking, we cannot prove either that it is or that it is not an iron ball when no one is looking. It was also George Berkeley who pointed out this unsolved philosophical problem.

What is certain, however, is that quantum physics can play no role in such mind games. It is based on an outdated version of the Copenhagen interpretation and its (correct) statement that quantum particles acquire specific properties only through observations or measurements (i.e., a measurement itself alters the system and projects it from a superposition of states into a definite state). For what exactly is meant by "observation" or "measurement" was left open in this statement. How does a measurement produce the concrete certainty of a single state from a system of probabilities? Is it only through the interaction with a macroscopic measuring instrument—as the Copenhagen interpretation originally assumed—or is it enough for a mouse to take a look, or for the system to interact with a single molecule of air? There is no reason to assume that this requires a human who consciously learns about the status of the quantum system through a physical measurement.

It is a huge leap from an electron which obtains certain properties through a measurement to the macroscopic objects of our everyday lives. From verifiable correct statements such as "Only the measurement determines the state of a quantum particle" it does not necessarily follow that "Only an observation determines the state of the world" (or, depending on what one is currently dealing with, the state of my health or that of my best friend). Such statements contain great speculative ballast and lack any physical and logical legitimacy.

Incidentally, in the explanation given by Thomas Wörner, or Frido and Christine Mann, it also remains unclear where and exactly how the mental aspect of things comes into play as the substance upon which everything is based. Thus, they leap over two deep intellectual trenches in one go: from the coherent and empirically validated measuring process of micro particles to a mystical worldview of the macro world of our everyday lives, and thence immediately from matter to mind and consciousness.

> Anyone who today appeals to quantum physics for the clarification of basic spiritual questions, such as the nature of the mind, is fishing in murky and speculative waters.

The Tao of Physics—The Quantum Esoteric Movement

Laymen who abuse quantum physics to support their obscure worldviews often move within esoteric circles or in dangerous proximity of such. Claims like "Everything is connected with everything" make members' hearts beat faster, all of them filled with such a desire for mysticism. The intellectual godfather of *quantum mysticism* uses the phenomenon of entanglement: "Quantum particles that are far apart from each other can be physically interconnected (entangled)." This statement then becomes: "We are all connected with each other, and also to the whole universe." This is, to put it bluntly, nonsense, and a prime example of how spirituality and science *should not* be combined.

Advertisements and assurances like "Quantum Healing: Immediately—and anyone can learn it!" or "Towards a Conscious Lifestyle with the Power of the Quantum"—suggest that engaging with quantum matters has become a thriving business in recent years. Great promises are made: at last, it is possible to reveal "exciting parallels between spirituality, medicine, and quantum physics and make those known to a wider audience worldwide." Such claims have become commonplace. There is talk of nothing less than miraculous healing and other fascinating things, with which someone can, as desired, cleanse their soul, set up their apartment (the keyword here is quantum Feng Shui), or with the help of "quantum resonance", find the perfect romantic relationship.

> The esoteric scene feels closely connected to quantum physics. The alternative and spiritual scenes swear by the word "quantum", using it to support all kinds of baloney.

The esoteric "post-physical quantum movement," as I would like to call it, looks back on a decade-long tradition. In the 1970s, the physicist Fritjof Capra wrote a book called *The Tao of Physics*, in which he claimed that the ancient mysticism of India is based on nothing less than the findings of modern quantum theory, albeit packaged in a poetic metaphysical form.[4]

The ideas set forth by Capra fell on fertile ground, and his book became the new bible for all those who wanted nothing more than to bring back spirituality to a "world-view disenchanted by scientific rationality" (Max Weber). But what form of spiritual experience can reasonably refer to quantum physics? Those who understand anything about quantum physics have a hard time finding a direct connection between the two.

> The vast majority of quantum effects take place far removed from our everyday lives, on tiny distance scales and unimaginably short time scales, and certainly not in the realms of human spirituality.

---

[4]F. Capra, *The Tao of Physics—An Exploration of the Parallels Between Modern Physics and Eastern Mysticism*, Shambhala Publications (1975).

This does not mean that quantum physics does not have exciting philosophical and conceptual implications, some even undermining traditional metaphysical views, as we have seen throughout the third part of this book. The reader will also remember Chap. 17, where we discussed Nāgārjuna and his rejection of any independent substance, of either a mental or a material nature. There are indeed some interesting connections between quantum physics and Buddhist teachings. The connection between a quantum particle and its environment and the abolition of subject–object duality are insights gained from modern physics, which can be compared to certain millennia-old ideas of Buddhist Mādhyamaka philosophy. Capra and others are quite right to point that out. Many of today's esotericists, however, are concerned neither with quantum physics nor with Nāgārjuna (and probably not with Capra, either).

Instead, we find websites that read like this:

> The foundation of the MatrixPower® method is an extract of the best-known quantum healing and transformation methods, Russian consciousness methods, ancient wisdom teachings, over 20 years of our own experience, the latest in quantum physics, and our medial and healing abilities. (…) Possible disturbances in one's own energy field are transformed, the pure consciousness opens up, and completely new ways can show thereby themselves. In this way you can specifically focus on topics such as health, partnership, career, cash flow, etc.[5]

Such nonsense is an important element of the market for esotericism, which in Germany alone generates revenues of 20–25 billion euros a year.

In conclusion, the "quantum" provides the perfect reference for a great deal of nonsense that cannot be easily repudiated, but at the same time lacks any real foundation. Who could contradict statements like the one above, given that so few people are really familiar with the topic of quantum physics?

> "Quantum esotericists" rarely have to justify their own ignorance. Hardly anyone dares to disagree when they make references to quantum physics. That provides a perfect framework for any esoteric mix for those too lazy to think more deeply.

---

[5]http://matrix-power.de/matrix-power/methode/index.htm (translated from the German webpage).

The Fallacy of Quantum Mysticism

That quantum physics is well suited for nonsensical esoteric ideas or fantasies about mind over matter is, of course, because the world of quantum particles does have bizarre surprises in store and in many ways does seem genuinely unusual and strange.

> In the past one said "God moves in mysterious ways...", if one did not know what to make of things. Today, many have turned his statement into "Quantum theory shows that ....".

The "quantum" satisfies many people's longing for a simplest possible, fundamental, universal, and binding worldview that we can invoke again and again without bothering to study its details. Knowing that quantum physics makes bizarre-sounding statements, the quantum esotericists cheerfully conclude that everything bizarre-sounding has to be part of quantum physics.

"Quantum esotericism" is supposed to quench our thirst for a simple worldview. To the advantage of its protagonists, because the quantum world is very abstract and difficult to understand, they can invoke it again and again, without being required to put their statements to the test. Who would challenge those who refer to the "quantum"? Physicists are not likely to belong to any esoteric movements. Everything related to "quantum" is wonderfully easy to use to exploit people who are on a spiritual quest. One might say that this would never harm anyone, except perhaps financially. But when diseases such as cancer are described to someone seeking advice as a "blocked quantum energy" and the sick person fails to be provided possibly helpful medical care, it can become life-threatening.

> Quantum physics may be weird, but not everything weird is quantum physics.

Quantum physics and quantum esotericism therefore have no common interface or overlap that would withstand a detailed and honest examination. Things look different when we come to consider quantum physics and *faith*, as the following chapter will show.

20

Quantum Physics and Faith: Explaining the Inexplicable

What kind of people were the pioneers of quantum theory? Did they devote themselves exclusively to the scientific method and rational thought? Were they as unworldly as Sheldon Cooper in the TV series *Big Bang Theory*? Quite the opposite! From many sources we know that their intense work on quantum theory forced them to carry out an exhaustive examination of their conceptual worldview, including questions of faith.

Faith, intuition, a sense of sublimity, the feeling of a unity of nature, pure amazement in the face of its miracles—for the pioneers of modern physics such sensations were inseparably connected with the scientific process of knowledge acquisition. In addition to formulas and laws, they were also concerned with questions of worldview. Their reflections and statements repeatedly allow us an insight into their worldviews (and in some cases, religiosity). These had their roots beyond all experience and comprehensibility, and thus beyond physics itself.

> For Planck, Einstein, Bohr, Heisenberg, and their colleagues, the search for knowledge was always also a spiritual search for a deeper meaning in nature.

The grandfather of quantum theory, Max Planck, formulated with these words:

© Springer Nature Switzerland AG 2018
L. Jaeger, *The Second Quantum Revolution*, https://doi.org/10.1007/978-3-319-98824-5_20

… the starting point of research is not in sensory perceptions alone and even science cannot do without a certain dose of metaphysics.[1]

Planck, born in 1858, was a deeply religious man; for him it was beyond question who or what was at the centre of this metaphysics:

For religion, God is the starting point, and to natural science he is the goal of every thought process.[2]

Einstein and the Dice-Throwing God

When it comes to quantum physics and religion, Albert Einstein's name inevitably comes up. Like Planck, he had grown up at a time when Newtonian physics ruled supreme. In Einstein's (and Newton's) worldview, everything could be calculated, including things like which side a tossed coin would come to lie. We only need to know all the parameters: the force with which the coin is thrown, the angle of the wrist, the friction of the movement against the air, etc. So there is no intrinsic randomness, it is just that the conditions are not all perfectly known. When physicists began to grasp the ideological meaning of quantum physics, Einstein did not want to give up the view that things really exist as we experience or measure them and that they evolve according to well defined and deterministic laws:

Quantum mechanics is certainly imposing. But an inner voice tells me that it is not yet the real thing. The theory says a lot, but does not really bring us closer to the secret of the 'Old One.' I, at any rate, am convinced that He is not playing at dice.[3]

Einstein could not believe that, as the experiments in the field of quantum physics suggested, it was only a spontaneous observer who would determine the state of a system. If that were true, he considered that the world would be inherently unpredictable. And this simply did not fit into his worldview.

---

[1]Translated from M. Planck: "Physikalische Gesetzlichkeit im Lichte neuerer Forschung" (1926), in *Vorträge und Erinnerungen*, Stuttgart 1949, p. 205.

[2]Translated form M. Planck, "Religion und Naturwissenschaft" (1937), in: *Vorträge und Erinnerungen*, Stuttgart 1949, p. 331; http://psychomedizin.com/medien/pdf/max-planck.pdf.

[3]Letter from Einstein to Max Born, December 4 1926.

Throughout his life, Einstein therefore remained a bitter opponent of the Copenhagen interpretation of quantum mechanics.

It is noteworthy that Einstein speaks of an "inner voice" and thus consciously moves away from a scientific line of argument. He was unable to physically justify his *faith* in the idea that there is no coincidence in physics. However, the "Old One" he is referring to is not the God of the Bible. Einstein shows how he imagined God in his response to a telegram from the New York rabbi Herbert S. Goldstein: "do you believe in god stop paid answer fifty words". Einstein replied in the same way on April 24, 1929 (with 29 words):

> I believe in Spinoza's God who reveals himself in the orderly harmony of what exists, not in a God who concerns himself with fates and actions of human beings.[4]

Einstein did not believe in a personal creator god. Shortly before his death he thus wrote to the philosopher Eric Gutkind:

> The word God is for me nothing more than the expression and product of human weakness, the Bible a collection of honourable, but still purely primitive, legends.[5]

Einstein was rather filled with a kind of "cosmic religiosity," consciously referring to Baruch Spinoza, the seventeenth century Dutch–Portuguese philosopher.

> In Einstein's worldview, God stands for an active principle in nature. It ensures that the world follows deterministic, rigorous causal laws that eliminate the need for intervention or control by God.

This belief was the basis of Einstein's religiosity, which gave him a deep reverence for the complexity and beauty of the world. He formulated his creed particularly clearly in 1949:

---

[4]Published under the title "Einstein believes in Spinoza's God", in *New York Times* 25. April 1929; W. Isaacson, *Einstein: His Life and Universe*, New York (2008), p. 388.

[5]Cited after B. Hoffmann, H. Dukas, *Albert Einstein. The Human Side*, Princeton, New Jersey (1981), S. 43.

The most beautiful experience we can have is the mysterious [...] A knowledge of the existence of something we cannot penetrate, our perceptions of the profoundest reason and the most radiant beauty, which only in their most primitive forms are accessible to our minds: it is this knowledge and this emotion that constitute true religiosity. In this sense, and only this sense, I am a deeply religious man.[6]

For Einstein, religion and science stood in a close relationship to each other, a relationship that found its expression in one of his most famous quotes: "Science without religion is lame, religion without science blind."[7]

Bohr and Far Eastern Thinking

But not only Albert Einstein thought about the bigger picture. The other pioneers of quantum theory also dealt with questions of worldview and spirituality, including Niels Bohr, who challenged Einstein in a fierce debate on philosophical issues. In the Bohr–Einstein debate, Bohr took the position of an agnostic. For Bohr, the world was just there, reality was just as it is, and we humans are an inseparable part of it. God—even Spinoza's God, who does not interfere with events in the world he has himself created—should play no role in the description of nature.

To understand the conceptual and philosophical consequences of the new atomic theory, Bohr preferred rather to turn to the findings of the great Far Eastern thinkers. After all, these had very early on dealt with the question of whether things are based on something substantial and permanently fixed, and how we humans relate to the phenomena we observe.

For a parallel to the lesson of atomic theory .... [we must turn] to those kinds of epistemological problems with which already thinkers like the Buddha and Lao Tzu have been confronted, when trying to harmonize our position as spectators and actors in the great drama of existence.[8]

---

[6]A. Einstein: *The World as I see It*, Citadel Press, New York (2008); original title: *Mein Weltbild* (1931).

[7]This is a modification of a formulation of Kant: "Thoughts without content are empty, intuitions without concepts are blind" (or "concepts without intuitions are empty, intuitions without concepts are blind)", I. Kant, *Critique of Pure Reason*, B75, A48.

[8]Speech on quantum theory in October 1937 on the occasion of the commemoration of the 200th birthday of Luigi Galvani (Celebrazione del Secondo Centenario della Nascita di Luigi Galvani) in Bologna, Italy. See also Niels Bohr, *Atomic Physics and Human Knowledge*, edited by John Wiley and Sons, New York (1958), pp. 19/20.

In Far Eastern spiritual traditions, Niels Bohr found more satisfying answers than in the Middle Eastern–Western religions which referred to a Creator God.

The relationship between Far Eastern beliefs and modern physics is thus not a recent discovery. Like Bohr, other protagonists of quantum physics felt a similar spiritual proximity to Buddhism, including David Bohm and Carl-Friedrich von Weizsäcker.[9]

One of the elements of Bohr's thinking that he related to Far Eastern wisdom was his concept of "complementarity". As for a coin, there are always two sides and they may well be in conflict with each other. While the western worldview is only able to conceive one of the two sides, i.e., it only accepts an either–or (heads *or* tails), Far Eastern traditions of thinking also allow ambivalence (heads *and* tails). There is no longer only *one* truth. The opposite may *also* be true. For a physicist, this is only a surprising worldview at first glance.

For quantum physics requires different perspectives (or truths) to describe physical processes at the atomic level. A photon or electron can be considered both a particle and a wave. Both are true. Although these truths are mutually exclusive, both approaches are necessary components of a complete description of the microcosm. However, knowing that both truths apply, it is *ourselves* who must decide whether we want to measure the electron as a wave or as a particle. We cannot measure both qualities at the same time. As observing subjects, we are therefore actively involved in the way we encounter the phenomena of the world. That's exactly what Bohr meant by complementarity.

Bohr was convinced that, for every deep truth (for example, the electron is a particle) there is an opposite that also contains a deep truth (the electron is a wave).

In a lecture in 1954, Bohr recommended applying the notion of complementarity to theology and said that classic examples of this principle there are justice and generosity.[10] Believers are asked to exercise both at the same

---

[9]In the 1980s and 1990s, von Weizsäcker met several times with Tendzin Gyatsho, the 14th Dalai Lama. In their exchange of ideas, they recognized clear parallels between quantum physics and Buddhist teachings (see Chap. 17).

[10]N. Bohr, *Atomic Physics and Human Knowledge*, Chapman & Hall, New York (1958), p. 82.

time. But how is that supposed to work? Almost always, justice works at the expense of generosity (how else could a judge sentence a person to a few years in prison?). And generosity is not always fair (does one always offer one's help to those who need it most, or only to specific people or in specific circumstances?). Instead of losing ourselves in the impossibility of practising justice and generosity at the same time, we must choose one side of the coin each time, fully aware that the other side of the coin exists and has its *raison d'être*.

Bohr sought to transfer his principle of complementarity to many more areas, including biology, psychology, and philosophy. He also said that it would eventually also provide clarity on social issues such as the relationships between different human cultures. One could in fact consider his belief a plea for intercultural tolerance. Confronted with contradictory principles or traditions, it can read: "I am right and you are *equally* right". Last but not least, for Bohr, science and faith stand in a relationship of complementarity—the two can stand side by side without inflicting damage on each other.

What is important is that the principle of complementarity does not mean finding a compromise or building an average of two extremes. It leaves both extremes for what they are and appreciates them equally.

> According to Niels Bohr, the principle of complementarity was not only important for quantum physics, but could be applied universally.

By the way, Niels Bohr was not the first to introduce ambivalence into the worldview of a Western society. Since it was founded over 450 years ago, the Jesuit Order, ever loyal to Catholicism and the Pope, has held the view that we humans are constantly exposed to conflicts in which a mindset of "as well as" is to be applied. One example is the tension between trust in God and personal responsibility. The Jesuits are not concerned with finding a balance between these two life attitudes. Every day, they try to come to terms with the fact that they *fully* trust in God and *at the same time feel 100% responsible* for their lives.

Heisenberg and the World of Values

A third pioneer of modern physics, Werner Heisenberg, also acknowledged the importance of spiritual questions. However, he avoided a direct answer to the question of whether he believed in God. He spoke (with reference to Einstein) of a possible "central order in things and events". But Heisenberg was far away from committing himself to a creator God. Nevertheless, citing Max Planck, who was one generation older than him, in the Christian religion he saw a pledge for human values that he considered to be important and to hold the same level of truth as scientific knowledge.

Spirituality and questions of value, Heisenberg said, are formulated in a language other than that of science, in a language much closer to art and images than to mathematics. While science makes statements about external reality, spirituality—or, in a narrower sense, religion—is about who we are and what we are here for, i.e., about the world of human values.

For Heisenberg and his colleagues, thinking about values was not just purely theoretical. Some of them took part in the effort to build the bombs that eventually fell on Hiroshima and Nagasaki. Heisenberg himself had been chairman of the German Uranverein, the Nazi project for the construction of a German nuclear bomb. How strongly he had been committed to this goal is still unclear among science historians today. What is certain is that he did not build a German bomb. But this was probably more because it was technically too difficult for his team to build rather than him wanting to absolutely prevent Hitler from getting it.

> For Werner Heisenberg, the values that describe what *should be* are at least as important as scientific knowledge, which describes what *is*.

Ideally, according to Heisenberg, spirituality or faith on the one hand and science, the objectively experienced apprehension of the world, on the other hand should not be in conflict with each other, but rather complement each other. In this view, Heisenberg came close to the complementarity principle of his teacher Niels Bohr.

With the principle of complementarity which allows two contradictory statements to be true at the same time, quantum physicists (primarily Bohr and Heisenberg) identified new aspects of spiritual questions. In a worldview that allows ambivalences, two truths A and B can apply 100%. The trick lies in the conscious decision, which view is "on" at any particular moment. Quantum physics was the inspiration for this new way of thinking.

> Within the worldviews of the pioneers of quantum physics, ambivalence became part of scientific thought. The dissolution of polarities combined quantum physics with some elements of Far Eastern thought.

Quantum physics was not only the trigger for complementarity to enter the minds of Westerners. It was also, as the reader knows from Chap. 16, the starting point for an equally dramatic change in physics: the disappearance of claims about absolute truth.

How Quantum Physics Pushed Faith to the Side

In Newton's time, physicists still believed that they could sneak a look at God's cards. Einstein also wanted to know "how God created the world". He was thereby referring to universal principles of nature, which humans only had to find, just like children searching for Easter eggs.

The physicist Richard Feynman already belonged to a new generation. He summed up the new worldview of physics as follows:

> And as you develop more information in the sciences, it is not that you are finding out the truth, but that you are finding out that this or that is more or less likely.[11]

Today's physicists have adjusted themselves to the fact that there is no absolute and final truth, but that they can only approximate the truth without ever getting to it.

The moment a person resigns himself to never finding absolute certainty, but only approximations to the truth, he no longer needs God as a tower of certainty. Feynman was among those who no longer needed either God or certainty.

> God was always invented to explain mystery. God is always invented to explain those things that you do not understand.[12]

Modern cosmology also contributed to pushing God farther and farther into the background. According to today's standard cosmological theory, at the beginning of the Universe, the entire cosmos was concentrated on a tiny

---

[11]Public address at the 1955 Autumn Meeting of the National Academy of Science, *The Value of Science*, held at the CalTech Campus Nov 2, 1955; in R. Feynman, *The Pleasure of Finding Things Out—The Best Short Works by R.P. Feynman*, New York (1999).

[12]Interview with R. Feynman, published in *Superstrings: A Theory of Everything?* edited by Paul C. W. Davies und Julian R. Brown, Cambridge University Press (1988), p. 208.

speck that had the diameter of a billionth of a trillionth of a trillionth of a centimetre. This spot started to expand quickly 13.8 billion years ago in the Big Bang, and the Universe came into existence.

When the Belgian priest and cosmologist Georges Lemaître published the Big Bang hypothesis in 1931,[13] physicists initially rejected it all round. It had too much the taste of a divine creation from the Bible to them. At that time, founding principles that referred to a divine creator were already banished from physics.

> Today, whoever argues like Einstein did by stating "God does not play dice" would risk being side-lined in the scientific community. Divine principles are no longer mentioned in physics, let alone discussed.

Not only has God disappeared from scientific dialogue, but all things considered, today's physics no longer serves *any* transcendent spiritual questions. There is no longer room in it for religious principles of creation or the definition of a fundamental substance in nature. For most physicists, the matter is clear: quantum theory is the authoritative theory about the origin and laws of our Universe. And this is all we need.

What Physics Has to Leave Open

However, physicists are still tormented by many unanswered questions. One of these is the fine-tuning of the constants in the Standard Model of elementary particles. It seems that the natural constants in the Universe are chosen exactly so that life is possible. Some physicists refer to the "anthropic principle", which simply states that we would not be around if the constants had taken different values. But, of course, this could not in itself provide a concrete explanation of *why* they take precisely these values, since such an argument would be circular.

At this point, scientists do not try to go further. They leave the field to spiritual traditions of thought, including religious belief, mainly because science cannot serve this kind of spiritual inquiry. Strictly speaking, science cannot even ask "why" questions, for these go beyond its methodological

---

[13]G. Lemaître, *The Beginning of the World from the Point of View of Quantum Theory*, Nature, 127, 706 (9 May 1931). A fist version of this work was already published in French in 1927.

framework. Of course, the fact that physicists as scientists no longer make spiritual references does not mean that they do not do so as humans. As *physicists*, they ask about mechanisms of action and causality in the world, but as *humans*, they also ask who we are and what we are here for.

> In the last 80 years, not only faith but any spirituality of any kind has disappeared from the scientific dialogue. For most physicists, questions that go beyond "what?" and "how?" have become a purely private matter.

Physics has thus been decoupled from religion. Conversely, the representatives of organized religions continue in their efforts to use the findings of physics to legitimize their beliefs. For example, many believers regard the Big Bang as a creation, triggered by God. In 1952 Pope Pius XII declared that the Big Bang theory was in deep harmony with Christian dogma and underpins the existence of a creator (even though there is nothing in the Bible about a time frame of 13.8 billion years). The anthropic principle is also seen by religions as evidence for the actions of a divine, ordering force in the Universe.

> The questions modern physics cannot (yet?) answer leave religious beliefs some basis for their creation myths.

Although he did not believe in a personal God, Einstein indicated that he would never try to fight faith:

> It is a different question whether belief in a personal God should be contested.. [...] For such a belief seems to me preferable to the lack of any transcendental outlook of life.

By this Einstein hardly meant that we should cling to a dogmatic belief, which he considered to disable one's own thinking, but rather a personal interest in the great philosophical questions. Without a worldview and without deliberate justification for one's own moral actions—that is, without dealing with spiritual questions, including the possibility of transcendent justification—man deprives himself of what makes him such a special being. Some people find the answers in a religion (whatever it may be), others want to approach the questions of worldview and the right way of acting on their own.

And so continues the quote from Einstein:

... and I wonder whether one can ever successfully render to the majority of mankind a more sublime means in order to satisfy its metaphysical needs.[14]

> It probably does not matter whether we believe in a divine power or not. What matters is that the world is so incredibly wonderful and we will probably never get enough of thinking about it.

In the next part of the book, after these excursions into the philosophical, spiritual, esoteric, and religious, aspects of quantum physics, we will return to the open physical problem of quantum theory that their creators so fiercely fought over. Here we get to the heart of all its possible interpretations, and at the same time, the basis for its future technological applications: the phenomenon of entanglement.

---

[14]Letter to Eduard Büsching (Oct 29, 1929), after Büsching sent Einstein a copy of his book; here cited from M. Jammer, *Einstein and Religion: physics and theology*, Princeton: Princeton University Press (2002), p. 51.

Part V

Entanglement—Getting to the Crux of the Matter

21

Cat Destinies: The Quantum Physical Measurement Problem

In 1935 Erwin Schrödinger wrote an essay that caused an intellectual uprising in the physics community. It bore the title *The current situation in quantum mechanics* (original German title: *Die gegenwärtige Situation in der Quantenmechanik*). In it Schrödinger asked: What do we actually measure in quantum systems?

As simple as this question may seem at first glance, it summed up the central problem of the Copenhagen interpretation, which the reader has already encountered on several occasions in previous chapters:

- Quantum objects are fundamentally inseparable from their measurement—thus it is impossible to separate (measuring) subject and (measured) object.
- The experiments and considerations of Bohr and his colleagues had shown that quantum objects only acquire their properties through their interaction with an environment. It followed that only a (macroscopic) measuring system could give quantum objects existence and essence.
- If quantum objects have no objective properties, then they cannot be credited with independent existence or reality as such.
- At the same time the existence of a measurement system is assumed a priori, which essentially implies an ontological separation of the world into micro and macro systems.

© Springer Nature Switzerland AG 2018
L. Jaeger, *The Second Quantum Revolution*, https://doi.org/10.1007/978-3-319-98824-5_21

The micro or quantum world manifests itself very differently than the macro world of our experience. But how can that be? And which world is the "right one"?

The answer of Heisenberg and Bohr in the context of the Copenhagen interpretation was pragmatic: quantum objects and macroscopic objects exist in two different spheres, each with its own laws—in the micro world quantum laws apply, and in the macro world classical physics. The two worlds are separated somewhere by the Heisenberg cut, where the laws of one somehow turn into the laws of the other. However, "somewhere" and "somehow" are rather unsatisfactory words for physicists.

It was known that the objectively indefinite nature of quantum objects gets somehow fixed only upon observation (measurement). Schrödinger's question—*What do we actually measure in quantum systems?*—was thus extremely challenging.

Crossing a Border

Schrödinger's essay dealt with this so-called *measurement problem*. He argued that an arbitrary separation into micro and macro worlds cannot be a satisfying solution, if only because states that can in theory exist only in the micro world can sneak into the macro world. Particularly famous became his thought experiment about a cat (fifth section in the article *Die gegenwärtige Situation in der Quantenmechanik*):

One can even set up quite ridiculous cases. A cat is penned up in a steel chamber, along with the following device (which must be secured against direct interference by the cat): in a Geiger counter, there is a tiny bit of radioactive substance, so small, that perhaps in the course of the hour one of the atoms decays, but also, with equal probability, perhaps none; if it happens, the counter tube discharges and through a relay releases a hammer that shatters a small flask of hydrocyanic acid. If one has left this entire system to itself for an hour, one would say that the cat still lives if meanwhile no atom has decayed. The first atomic decay would have poisoned it. The psi-function of the entire system would express this by having in it the living and dead cat (pardon the expression) mixed or smeared out in equal parts. It is typical of these cases that an indeterminacy originally restricted to the atomic domain becomes transformed into macroscopic indeterminacy, which can then be resolved by direct observation. That prevents us from so naively accepting as valid a "blurred model" for representing reality.

The genius of Schrödinger's thought experiment is that it links the micro and macro worlds. Only a single atomic nucleus has to decay and give off radiation, which then has a decisive influence on a macroscopic object—a cat.

As long as no measurement takes place, the nucleus is in a state of superposition of "decay" and "not decay". Only by measuring, i.e., opening the door, does the wave function of the atomic nucleus collapse and its objectively indefinite state become a definite state.

The fact that the quantum laws allow a superposition of states does not seem wholly unreasonable, provided that we keep it to the microcosm! But the causal chain constructed by Schrödinger relates events of the microcosm directly with those of the macrocosm, and hence the world of our direct experience. The experimental design not only determines the state of the atomic nucleus, but also the state of the cat in the steel chamber. This consists of a superposition of "dead" and "alive"; the cat is objectively in both states at the same time, as long as the door to the box is not opened, so that the measuring process has not yet been carried out. This contradicts our notion of everyday realism, according to which the cat must be either dead or alive, regardless of whether we look at it or not. It was precisely this paradox that Schrödinger wished to bring out.

Schrodinger's thought experiment showed that basically nobody knew where to place Heisenberg's cut, which separates the laws of the microcosm from those of the macrocosm.

Discussion on the State of Things

Schrödinger's cat forced physicists to position themselves. Nobody doubted that the quantum laws were valid in the micro world. Calculations and predictions from quantum theory coincided too well with measurement. The question was: How far into the macro world does its power reach? Where in the chain

- atomic nucleus that decays or does not decay
- Geiger counter
- poison bottle
- cat
- observer

do we have to apply the Heisenberg cut? Where exactly does the boundary lie between the microcosm with its quantum nature and the subjective conditionality of objects and the macrocosm? Where can we trust our intuition and separate objects from subjects?

If the dividing line runs between the atomic nucleus and the Geiger counter, there is no cat in superposition. As part of the macro world, it is objective, i.e., it is dead or alive regardless of our observation. However, if the dividing line lies at this point, then the Geiger counter and the atomic nucleus form an inseparable structure consisting of object and measuring system; the processes that cause the Geiger counter to click are clearly taking place in the micro world and are subject to quantum laws. So the cut would have to be placed further in the direction of the poison bottle, the cat, and the observer.

But where? If atomic nuclei and Geiger counters can be linked to form *one* wave function ("form an inseparable structure"), this also applies to all higher-level systems, including the poison bottle. Maybe the cat is also part of the measuring system? That is the life-and-death issue (for the cat). If the cat were a part of the measuring apparatus in the sense of quantum physics, it would actually be as impossible for the cat as for the atomic nucleus to say that it is in an objectively determined state before the measurement has been carried out.

Only the following is certain:

- above the Heisenberg cut is the measuring system with its macroscopic features, which assigns unique properties to the quantum objects. Any level above contains systems that contribute nothing more to the measurement, and things obtain an objective existence.
- Below the cut, objects do not possess any independent existence of their own.

Nowhere in the chain from the atomic nucleus to the observer does there seem to be a suitable place for the Heisenberg cut.

Entanglement Enters the Stage

Different physicists placed the concrete position of the Heisenberg cut at quite different points within the chain quantum object–measuring apparatus–observer:

- Born and Pauli suspected it to be right after the quantum object.
- Bohr in the first *macroscopic* part of the measuring apparatus, i.e., the Geiger counter.
- von Neumann between the first measuring device and the observer.
- Eugene Paul Wigner, a Hungarian born American physicist we will learn more about in the next chapter, even positioned it *in* the observer.
- Heisenberg said that the exact position of the dividing line could be chosen arbitrarily, because the mathematical description makes the same experimental predictions no matter where the cut is made[1]:

It has been shown that in our exploration of atomic processes a peculiar dichotomy is unavoidable. This dichotomy gives rise to the need upon the description of atomic processes to draw a line between the measuring apparatuses of the observer, described by classical concepts, and the observation object whose behaviour is represented by a wave function [...]

The dividing line between the system to be observed and the measurement device is directly defined by the nature of the problem, but for obvious reasons cannot bring about a discontinuity in the physical process. For that reason, within borders, there must be complete freedom in choosing the position of the dividing line.

Schrödinger's thought experiment set in motion a fruitful discussion among physicists. There were good reasons for every position to be the location of the Heisenberg cut, but there were also good reasons against all of them.

Schrödinger himself also took a clear position regarding the measurement problem. He located the cut beyond the cat, and this for a particular reason. In the same article in which he described his thought experiment (§ 10), he introduced a concept that still shapes the discussion on quantum physics to this day and at the same time constitutes the basis of today's hopes for completely new quantum technologies: the concept of *entanglement*.

Entanglement describes the quantum physical fact that multi-particle systems of two or more particles can no longer be described as a combination

---

[1]W.Heisenberg, *Wandlungen der Grundlagen der exakten Naturwissenschaft in jüngster Zeit*, Talk at the German Society of Scientists and Medical Doctors (Gesellschaft deutscher Naturforscher und Ärzte), Hannover, 17. September 1934, Angewandte Chemie 47 (1934).

of independent one-particle states, but only as a common state, which in principle must be described *within a single wave function*. In this sense, according to Schrödinger, macroscopic measuring systems should also be connected to quantum objects. Rather casually, he wrote:

> According to the compulsory law of the total Ψ function, it [the Ψ function of the object of measurement] has been tangled up with that of the measuring instrument (…)

He then describes the central problem of entangled quantum systems (§ 15):

> Best possible knowledge of a whole does **not** include the best possible knowledge of its parts. [emphasis by Schrödinger]

What Schrödinger meant by this is that one cannot simply separate the entire wave function into its parts. Their parts are interwoven (entangled). For Schrödinger this meant that the entire system of atomic nucleus–Geiger counter–poison bottle–cat must be described in *a single* wave function. So the cat is indeed in a state of superposition as long as no one is looking to see if it is alive or dead.

> For Schrödinger, all the components of his thought experiment, from atomic nucleus to cat, are in fact entangled. Thus it was clear to him that the quantum laws also apply in the macro world.

In the following years and decades, the idea of entanglement brought into play by Schrödinger proved to be the crux of the matter in the quantum world. Even in 1935, he wrote:

> I would not call that property [entanglement] **one** but rather **the** characteristic trait of quantum mechanics, the one that enforces its entire departure from classical lines of thought [emphasis by Schrödinger].[2]

He realized that this subtle property corresponds to nothing other than the fact that the wave function of several particles can no longer be factorized

---

[2]E. Schrödinger, *Discussion of Probability Relations between separate systems*, Proceedings of the Cambridge Physical Society, 31, 55 (1935).

into the wave function of individual particles. Chapter 26 will show that Schrödinger's ideas back in 1935 are quite close to the modern interpretation of the measurement problem. But before that, physicists had to take a few detours.

Everything Is True

The most radical solution proposed for the quantum mechanical measurement problem was initiated by the American physicist Hugh Everett. In 1957 Everett addressed the measurement problem by asking on which levels of a system we can and cannot speak about realities. If the Heisenberg cut, as proposed by Schrödinger, lies beyond the cat, the cat possesses neither independent existence nor objective reality. As this contradicts our everyday experience, Everett simply attributed a physical reality to *all* possible states. So there exists one live cat *and* one dead cat. How does that work?

According to Everett, as soon as the lid is opened, two parallel universes are created (in fact, depending on the quantum system, many universes). In one the atom has decayed, in the other it has decayed. In the first universe we thus find the cat dead when we open the box, and in the second universe we find it alive. Instead of one universe there exist innumerable universes, together forming the so-called multiverse. Each time a measurement is performed on a quantum object, the multiverse separates into further universes in each of which a particular state is realized. What we experience as reality in "our" universe corresponds to just *one* of countless possible stories.

> The many-worlds theory turns every potentiality of a quantum system into a complete reality. It replaces the single unique universe of our intuition by a multiverse made up of countless universes.

Everett's solution to the measurement problem presupposes that there are myriads of different worlds instead of a single world. Despite this rather fantasy-like notion, there exist several good arguments for the many-worlds theory:

- It solves the measurement problem, since the parallel existence of possible cat states is no longer a contradiction. Both states—dead and alive—have become real in their own universes.

- It resolves the objectivity problem. In the multiverse, quantum objects are no longer indeterminate, but have their own reality in an objective way. In every single universe each particle exists in a fixed, clearly defined state. Every world on its own is a realization of a certain state of its quantum objects.
- Since all state possibilities become real, one no longer has to speak "only" of their probabilities. In the micro world, as in the macro world, things are deterministic.

Apart from the fact that the many-worlds theory seemed too far-fetched for most physicists, there is another catch to it: it cannot be proven, because it contains the statement that, from the vantage point of any given universe, every other universe is in principle unobservable. Thus, Everett's theory is empirically non-verifiable (and probably not even falsifiable). For this reason it is strictly speaking not a scientific theory, but resembles rather a statement from mediaeval scholasticism.

> Everett's many-world theory will never be proven, nor will it ever be refuted.

The Final Question

Let us return to Heisenberg's cut. An interesting approach to solving the measurement problem is to ask (as Schrödinger asked): Does the Heisenberg cut even exist? Without it, quantum physics would be universally valid; every system would be connected to the next higher measurement system (Geiger counter, poison surface, cat, observer, etc.) in a single ever growing wave function. This chain could be continued further and further without limit and the measurement would never be completed. Ultimately, the description of the measuring process would have to be applied to a *single* wave function representing the entire universe. But then who would be the external observer that, through his act of observation, "brings forth" the existence of the universe?

Another idea is to question the impermeability of the Heisenberg cut: Is the separation between the macro and micro worlds really as sharp as Heisenberg and Bohr assumed? In fact, there is some evidence that the limit cannot be absolute, as it has long been considered. For we now know that, that under suitable conditions, quantum effects can also occur in the macro world (e.g., Bose–Einstein condensation, laser effect, superconductivity, superfluidity).

Superficially, the measurement problem consists of the question of where the Heisenberg cut is to be applied, but looking more closely, it requires much more fundamental considerations about the way the micro and macro worlds interact.

As of the mid-1930s, physicists began to study the measurement problem more intensely. This was not just for philosophical reasons, but in fact mainly for practical purposes, because as long as the measurement problem was not understood, processes in the micro world were not entirely understood either. In the discussion about the measurement problem all the wrong and the right steps along the path towards a refined and ultimately consistent quantum theory can be made explicit.

The measurement problem is so central to quantum physics that the next five chapters will revolve around this topic. It will become clear that:

- there is no definite boundary between the microcosm and macrocosm,
- the bizarre properties of quantum physics are by no means confined to the microcosm,
- rather, there is also a lot of quantum physics in the macro world,
- this macroscopic quantum physics enables the development of entirely new technologies,
- consciousness and the human mind play no role in quantum physics.

The measurement problem opens the door not only to a philosophically conclusive interpretation of quantum theory, but also to a more comprehensive physical understanding of the quantum world—and thus to exciting new technological applications.

22

Wigner's Friend: Quantum Physics and Consciousness

A thought experiment called "Wigner's friend" illustrates just how far the game with Schrödinger's cat can be played. It was first formulated by the Hungarian–American physicist Eugene Wigner, who worked on the mathematical and theoretical foundations of quantum physics, and hence also on the measurement problem.

In 1961, when he formulated it, it was clear to physicists that, if a quantum mechanical system is in a state of superposition before measurement, the measurement dissolves this indistinct state, due to the collapse of the wave function that accompanies it, and exact, discrete measured value results. Like some other physicists at the time, Wigner saw a way out of the problem of measurement by giving this property-determining role to human consciousness. The human senses determine that the state of the system can ultimately be clearly identified and this ends the superposition of quantum states. For Wigner, the non-material consciousness of the observer constituted the boundary between quantum mechanics and classical mechanics. He wrote[1]:

> It was not possible to formulate the laws of quantum mechanics in a fully consistent way without reference to the consciousness.

---

[1]E. Wigner, Remarks on the Mind–Body Question. In: Wheeler, Zurek (1983). S. 168–181. (Original in: I. J. Good Hrsg.): *The Scientist Speculates*, London, S. 284–302 (1961).

© Springer Nature Switzerland AG 2018
L. Jaeger, *The Second Quantum Revolution*, https://doi.org/10.1007/978-3-319-98824-5_22

For Wigner, it is only the human being, through his perception of the measurement result, who can end the theoretically infinite sequence of the measurement process.

Like all other physicists, however, Wigner had no idea how the measurement process could trigger the collapse of the wave function, nor how human consciousness could accomplish that.

The Third Man

Wigner realized, of course, that he had just launched a new paradox with his suggestion: What happens when at the end of the measurement chain two people are connected in series? His thought experiment, known as "Wigner's friend", introduced another human observer into the chain from the atomic nucleus to the Geiger counter, poison bottle, cat, and observer.[2] Wigner imagined himself watching from the door of the laboratory as a friend of his looks into the box containing Schrodinger's cat. By Wigner asking his friend whether the cat is alive or dead, the friend is no longer the final element of the measurement chain, but just another component.

If, instead of a friend, a lifeless measuring device were to examine the state of the cat, its state would also be in a superposition state with that of the cat, the poison, and the atomic nucleus. Provided that the Heisenberg cut actually occurs directly before the human consciousness that recognizes the result, the reduction of the state of the entire measuring system into a particular state will only take place *after* this new measuring device. From the point of view of the observer at the door, however, it makes no difference whether another lifeless measuring device is inserted into the chain of measurement or a human being.

The logical consequence is that all parts of the connected quantum physical system, from atomic nucleus to Geiger counter, poison bottle, and cat, *right up to and including Wigner's friend*, are now in an indeterminate state. His possible answers "I see a dead cat" and "I see a live cat" are also in a state superposition. Only when Wigner asks his friend what he sees in the box, the spell is broken, and from the superposition of indeterminate characteristics there arises a definite reality, this time in the world of Wigner's experience itself.

---

[2]In fact, Wigner's example was that of a photon in a superposition of two states, one of which produces a flash that may or may not be seen by his friend. His example is often set in the context of Schrödinger's cat, which is what we have done here.

Wigner extended the causal chain from the possibly decaying atom to the observer by including another human observer. As long as the person does not say what she sees, her consciousness will remain in an indeterminate state.

The uncertainty is only resolved when the subject—in this example, Wigner as external questioner—learns about the condition of the cat. Then there exists a clear account of the measured system state. The paradox of this logical extension of Schrödinger's thought experiment is that it is not until the moment when *I myself* learn about the state of the cat, whether from my friend or from a measuring device, that the reduction of the cat state superposition is completed. When we consider that there could be a third observer who asks Wigner ("myself") if the cat is alive, the thought experiment becomes completely absurd. For all three people, viz., Wigner's friend, *myself*, and also the third person, it is only with their own conscious knowledge of the cat's condition that the uncertainty is finally dispelled. Depending on the observer, the quantum-mechanical reduction thus takes place at different times.

In the Cartesian tradition, at least in one's own consciousness, the quantum mechanical indeterminacy has to end. "I think, therefore I am" becomes: "I think, so it's clear what actually happened to the cat."

In the 1970s, Wigner eventually rejected his idea of a reality created only by the subjective consciousness of a human being. Today the consciousness-based interpretation of the measurement problem finds little support in the physics community. It remains, however, a good example of the many attempts to map the unknown world of quanta into which physics had just launched itself.

Quantum Effects in Biological Systems

Interpreting consciousness as the conclusive step in the process of a quantum physical measurement is no longer viable today. For in the 1970s the concept of *decoherence* found a more convincing way to address the paradoxes in Schrödinger's and Wigner's thought experiments (for more on this, see Chap. 26). But the question that had led Wigner and Schrödinger to these thought experiments has not been fully answered to this day: Where on the upward scale to the macroscopic world do quantum effects cease to play a role?

One aspect of this question is clearly this: Do quantum effects also play a role in biological processes? It will turn out that this question ultimately leads yet again to a speculative connection between consciousness and quantum physics, but in a quite different way to what Wigner had originally imagined.

The development of modern molecular biology has given new meaning to the question of where to locate Heisenberg's cut. In 1944, in his seminal paper "What is life?",[3] Erwin Schrödinger already described possible connections between quantum physics and biology. In it he introduced, among other things, the idea of an "aperiodic crystal" containing genetic information in its configuration of chemical bonds. This thought can be understood as a first indication of the existence of something like DNA (whose structure was discovered nine years later by Francis Crick and James Watson).

Schrödinger also expressed the idea that the mutations observed in every living being, the basis of all earthly evolution, arise from quantum leaps in these crystals. In fact, in 1963 the Swedish physicist Per-Olov Löwdin discovered a quantum physical mechanism as an explanation for DNA mutations.[4] The genetic code is given by the arrangement of the bases adenine (A), guanine (G), cytosine (C), and thymine (T) in the DNA. The opposite base pairs of the double helix are connected by very loose chemical bonds called hydrogen bonds. Here, too, the laws of quantum physics are at work: individual protons of this complex giant molecule can change their place through the quantum mechanical tunnelling effect. Over time this results in a small but cumulative probability that single base pairs (A-T and G-C) will spontaneously transform into the tautomeric pairs A* -T* and G* -C*. These have a different pairing behaviour, so that at the next duplication a mistake will inevitably occur in the genetic base sequence—in other words, a mutation. Quantum effects therefore play a significant role in the processes through which life evolves, adapts, and conquers new niches.

---

[3]E. Schrödinger, "What Is Life? The Physical Aspect of the Living Cell." Lectures at the Dublin Institute for Advanced Studies at Trinity College, Dublin, in February 1943. Published in 1944 by Cambridge University Press.

[4]P.O. Löwdin, Proton Tunneling in DNA and its Biological Implications. *Reviews of Modern Physics* 35 (3), 724–732 (1963).

For Löwdin, his discovery was reason enough to speak of a new research field which he called *quantum biology*.

> The replication of DNA and RNA is much more error prone than environmental conditions such as exposure to UV radiation can explain. Löwdin showed for the first time that quantum effects can be responsible for mutations.

In addition to translation errors in the replication of DNA and RNA, biologists have discovered a whole series of biological processes in which quantum effects play a central role[5]:

- Photosynthesis: The photosynthesis of plants is the basis of all life on Earth. In the chlorophyll molecule, light energy is transformed into the excitation of electrons, and then converted into chemical energy for the cells. What is crucial in this process is to transfer the electron excitation efficiently and promptly to suitable locations within the cell. For a long time, it was not known how photosynthesis could achieve the observed transfer efficiency of more than 99%, a value that cannot be explained in the context of classical physics (for example, by diffusion). Only in 2007 was it proven that the quantum mechanical tunnel effect and entanglement come into play here.[6]
- Cellular respiration: As with photosynthesis, quantum tunnelling also plays an important role in enzymatic activity in the context of cellular respiration. Enzymes direct and control the chemical reactions of cellular oxidation, which produce the necessary energy for the cells. Again, quantum processes ensure that electrons are transmitted efficiently and quickly over long distances within the cell.

> In cell respiration and photosynthesis, the quantum effect of tunnelling ensures the necessary efficiency of these processes.

---

[5]See also: J. McFadden, J. Al-Khalili, *Life on the Edge: The Coming of Age of Quantum Biology*, Crown. Lake Arbor (2015).

[6]See also: G. Engel et al., Evidence for wavelike energy transfer through quantum coherence in photosynthetic systems. *Nature* 446, 782–786 (2007).

Besides these elementary contributions to life there are many other biological processes that would not work without quantum effects. Here are two more examples:

- Our sense of smell: For a long time, biologists explained our sense of smell with a "lock-and-key model", i.e., we smell a certain odour when the shapes of the odour molecule and the corresponding receptor in the nose match perfectly. However, it turned out that molecules with very similar forms can produce very different odour sensations. On the other hand, there are completely different molecules that trigger the same sense of smell. Biologists discovered that our nose does not perceive the shape of the molecules involved, but the specific frequency with which the chemical bonds between certain atoms vibrate. Once again, the tunnelling effect is responsible: inside the receptors in the nose, electrons tunnel through a barrier and trigger a stimulus. This happens only in the presence of a molecule with a very specific vibration frequency.
  In physics, this specific quantum phenomenon is referred to as "inelastic electron tunnelling". We could say our nose behaves like a scanning tunnelling microscope.
- Magnetic sense in birds: In Chap. 2 there was already mention of some birds' ability to orientate themselves using the terrestrial magnetic field. In 1976, the German biophysicist Klaus Schulten found a possible explanation for this ability in the form of the so-called radical pair mechanism. Here we give a more detailed description of the corresponding processes in birds' eyes.
  A cryptochrome molecule pair briefly activated by light alternates between two quantum-mechanically possible states that differ only by the spin of the electrons involved in them. The two states are thus in superposition, i.e., the electrons are entangled. After a short time, the pair disintegrates and, depending on which state it is in, it yields products with distinct properties. The trick here is that, due to the spin properties of the participating electrons, the final state is dependent on the inclination of the Earth's magnetic field; the birds thus recognize at what angle to the field lines of the magnetic field they are flying. (It has not yet been fully proven that it is cryptochrome molecules that are responsible for this effect, but the involvement of entangled electrons is almost certain.)

Quantum laws are not only important for the molecules that are responsible for life, like DNA, RNA, enzymes, chlorophyll, hemoglobin, etc. Many other forms and properties of life would be unthinkable without quantum effects.

Quantum Physics at the Beginning of Life

Some quantum biologists would even go so far as to say that quantum effects were instrumental in the genesis of living organisms. If their idea proves correct, this would be a major step towards answering one of the most important open questions of any science: the origin of life on Earth.[7]

So what exactly is the problem here? To understand the mechanisms that led to the creation of living organisms, we must ultimately explain how, through various intermediates, carbohydrates and amino acids were able to form the RNA and DNA which could then be used as carriers of the genetic code. The juxtaposition of random chains of individual amino acids is easy to accomplish by experiment. But the leap between such random chains and DNA or RNA sequences that self-replicate and encode biological information in their sequence is simply too large for their creation to be accidental.

A look at a chain of only 100 elements, with simple components like red and green beads, can serve to illustrate this. The number of possible arrangements in this simple system is 2^{100}, already far more structural variations than there are atoms in the entire universe! (For RNA and DNA, which consist of *four* different building blocks, this number is orders of magnitude higher than for the above system of 100 binary components.) A chain that can serve as a blueprint for all enzymes, and which in addition ensures that the chain can duplicate itself, with all this coming together within a few billion years by pure chance, would be far too much of a coincidence.

Today there still exists no clear and unambiguous explanation for the origin of complex information carriers such as DNA and RNA. The probability of their emerging by pure chance is close to zero.

Could it be that the rules of quantum physics worked in some way to launch the formation of the first self-replicating polymers? If the chain from the above example worked according to the principle of a quantum computer (see Chap. 4), its 100 components would be in a superposition of all possible configurations at the same time. In biological reality, one could imagine this to materialize as follows. A random, not self-replicating RNA-like molecule (a so-called protoenzyme) among others consists of numerous protons (hydrogen nuclei) and electrons. Through quantum tunnelling

---

[7]See also (in German): L. Jaeger, *Wissenschaft und Spiritualität*, Springer Spektrum, Heidelberg (2017), Chap. 5.

effects, these can each take on different states by the protons and electrons tunnelling unhindered through its structure. This creates superpositions of trillions and trillions of different configurations. In essence, all possible combinations are realized at the same time. In this "biological quantum computer", a unique state configuration able to replicate itself might arise from the vast set of all possible states.

> A quantum-computer-based molecule might have an efficient search strategy for finding exactly that state which could replicate itself and represent the genetic code.

Quantum Physics in Our Head

The origin of life, evolution, sensory perceptions … none of this would be possible without quantum effects. Could quantum physics perhaps even offer an explanation for the mysterious processes in our brain that lead to consciousness?

Wigner had concluded that our consciousness puts an end to superpositions and ultimately provides an indeterminate state with an objective and unambiguous status. The English mathematician and theoretical physicist Roger Penrose turned the tables: he said that it might not be that our consciousness determines what happens at the quantum level, but rather that our consciousness is itself only possible thanks to quantum effects. To avoid any idea that some nutty scientist has come up here with yet another crazy idea, let us remember that Penrose is one of the leading minds in mathematics and physics today and has been showered with prizes and honours for good reasons.

> Wigner had speculated that our consciousness might be what determines quantum processes. Penrose put things the other way around, suggesting that quantum processes may control our consciousness.

Penrose developed the daring idea that consciousness is in essence a quantum physical phenomenon in the 1990s, together with Stuart Hameroff, a professor of anesthesiology and psychology.[8] Central to their considerations

---

[8]R. Penrose, *The Emperor's New Mind: Concerning Computers, Minds, and the Laws of Physics*, Oxford University Press (1989); R. Penrose, *Shadows of the Mind: A Search for the Missing Science of Consciousness*, Oxford University Press (1996).

are the millions of tubular structures contained in all plant and animal cells, the so-called microtubules, made up of long chains of the protein tubulin.

In fact, tubulin tubes contain alternating elongated and contracted sections. That their sequences could correspond to a code has on the face of it nothing to do with quantum physics. But Penrose and Hameroff suggested that the different states of microtubules might be in a superposition. Entanglement of the tubulin tubes of one neuron with those of other neurons could essentially create a quantum structure looking like a molecular quantum computer. The two researchers estimate that by a controlled process of state reduction, the common wave function of the neurons would collapse about 40 times per second with the resulting unique configuration producing a conscious experience. This is consistent with the observation that human consciousness does not consist of an uninterrupted stream of impressions, but is broken up into certain time intervals.

> According to Penrose and Hameroff, microtubules do not (only) serve as a mechanical support for the cell, but function in our nerve cells like qubits in a quantum computer.

However, most brain researchers, biologists, and physicists are quite sceptical of the theory that our brain works like an entanglement-based quantum computer. For the quantum effect of entanglement is much more susceptible to interference than quantum tunnelling. According to current knowledge, microtubules and neurons are simply too large and too complex to remain in entangled states and function as qubits.

The fundamental objection to the quantum consciousness theory arises from the fact that, in the "wet and warm" brain, quantum coherence, as assumed by Penrose and Hameroff, could not be maintained long enough, even for very small systems (electrons, protons). In order to model the quantum physical effect of entanglement *in the laboratory*, physicists have to cool down the environment to very low temperatures and eliminate influences from other atoms as far as possible. In *living tissue*, with its comparatively high temperatures and water molecules swirling around more or less everywhere, quantum effects would be eliminated in next to no time.

On the other hand, it is quite possible that processes in our brain are based on other quantum effects, such as the tunnelling effect. Given their immense importance in processes such as photosynthesis and respiration, it would actually be rather surprising if they did not play a role in our brain.

For example, as early as the 1990s, brain researcher and Nobel laureate John Eccles speculated that quantum physics might be involved in the interaction of neurotransmitters and synapses.[9] Moreover, the processes around the ion channels in nerve cell membranes are another candidate for such quantum processes.[10] The search for quantum effects within the biological processes of our perception is a very active line of enquiry today in brain research and quantum biology.

> Even though no explicit processes based on quantum effects have so far been detected in the brain, a positive finding would by no means come as a surprise.

---

[9]J. Eccles, *How the Self Controls its Brain*, Berlin, (1994); See also: F. Beck, Synaptic Quantum Tunnelling in Brain Activity, *NeuroQuantology*, Vol. 6, 2 (2008).

[10]See M. Donald, *Quantum Theory and the Brain*, Proceedings of the Royal Society (London) Series A, Volume 427, S. 43 ff. (1990).

23

EPR and Hidden Variables: The Debate About Spooky Action at a Distance

The article in which Schrödinger introduced his famous thought experiment, and with it the concept of entanglement, did not come out of nowhere. It was the direct answer to a no less significant paper by Albert Einstein and his American colleagues Boris Podolsky and Nathan Rosen, published in the same year: *Can the quantum mechanical description of physical reality be considered complete?* Just like Schrödinger's, this paper would make a major contribution to the history of science.

> The aim of the article published by Einstein, Podolsky, and Rosen in 1935 was to use a paradoxical thought experiment to attack the Copenhagen interpretation of Niels Bohr and reduce it to absurdity.

According to Bohr (and most other quantum physicists) the following statements were supposed to hold true:

1. There is no objective, measurement-independent existence of quantum particles. The indeterminacy of the quantum mechanical wave function draws a *non-real* picture of the quantum world.
2. This missing physical reality means that in principle only statistical statements about the state of particles are possible before an actual measurement is performed. Thus, the quantum world is *indeterministic*.
3. During the measurement, the wave function collapses into a single state. For example, the impact of an electron on a photographic plate has the

© Springer Nature Switzerland AG 2018
L. Jaeger, *The Second Quantum Revolution*, https://doi.org/10.1007/978-3-319-98824-5_23

consequence that it no longer has wave properties, but must be described as a particle that has blackened a single point on the plate. The information that the electron has hit the plate as a particle at a certain location must reach all other locations of the collapsing wave without any time delay, otherwise a single electron could cause several black spots. (see also Chap. 8)

Einstein had never come to terms with the statement that the quantum world is non-real and undetermined. In addition, his thought experiment proved that the third statement necessarily leads to the conclusion that quantum physics, as interpreted by Niels Bohr, also has to be *non-local*.

> The Copenhagen interpretation describes the micro world as non-real, indeterministic, and nonlocal. Einstein and several other physicists did not believe this threefold break with classical physics would hold up.

Einstein Resists

So here is the ingeniously simple thought experiment by Einstein, Podolsky, and Rosen, later referred to as the *Einstein–Podolsky–Rosen paradox*, or EPR paradox for short. Two quantum particles interact and are thus described by a common wave function. Their total momentum is known, but the momentum of each particle is indeterminate; the particles are in a superposition of many momentum states. Now the two particles are allowed to move to two widely separated places. On one of the particles, a measurement of its momentum is performed. The following happens:

- The wave function collapses and a defined value can be assigned to the momentum of the measured particle.
- For the partner particle the common wave function also collapses, without any time delay (instantaneously). Its properties are thus no longer uncertain.
- Its momentum is now precisely defined; it must assume the unique value which corresponds to the law of momentum conservation.[1]

---

[1] Often the EPR paradox is explained by the example of two electrons and their spin, but the original version due to Einstein, Podolsky, and Rosen deals with two unspecified quantum particles and the measurement of their momenta.

The EPR paradox is this: by measuring one particle and determining its state, the state of the other particle, possibly far away from it, has to be equally determined, and this instantaneously. But how, over the distance and without information transfer, can the second particle instantaneously "know" that it must take on exactly the complementary value for its momentum?

> The EPR thought experiment put it in a nutshell: according to the Copenhagen interpretation, there must be an instantaneous, long-distance effect.

The representatives of the Copenhagen interpretation had so far only stated rather vaguely that, after measurement of the first particle, the second particle somehow "knows" what momentum it must possess. However, they were simply unable to explain the effect. Einstein, Rosen and Podolsky said that such a phenomenon was utterly impossible. There could be no immediate effect, because the theory of relativity formulated by Einstein himself ruled out the possibility that any information could travel faster than light. Such a "spooky action at a distance" would violate the basic principle of locality. Locality means that direct interactions are possible only between two systems in immediate proximity. If distances are to be bridged, there must be a transmission mechanism (for example, electromagnetic waves) and this would take time. Einstein concluded that the Copenhagen interpretation could not give a complete description of the quantum world.

> Einstein's argument went like this: if the Copenhagen interpretation is correct and the quantum world is not real, then it must necessarily be non-local. However, as such "spooky action at a distance" is impossible, there is something wrong with the Copenhagen interpretation.

But how did Einstein explain the experiments showing that quantum objects do not have unique properties before measurement (i.e., are not real)? He was convinced there had to be a mechanism that physicists had not yet discovered. The idea of such "hidden variables" has been haunting discussions in the physics community for years. Einstein finally took it to its logical conclusion. Such variables should determine the properties of the particles that are detected during the measurement *before the measurement*, but be non-measurable themselves. In his opinion, only such hidden variables would turn the so far incomplete quantum mechanics into a consistent theory.

The hidden variables would ensure that even in the micro world:

- there exists an objective physical reality,
- events are deterministic,
- the local character of physics is maintained.

These three points correspond to classical physics and *contradict* both the Heisenberg uncertainty principle and the Copenhagen interpretation, because these described the quantum world as non-real, non-deterministic, and non-local.

> The "hidden variables" should right from the start contain information about how the particles should behave at the time of the measurement. Thus, quantum physics would be a *real*, *deterministic*, and *local* theory.

Clash of Views

What did the events look like from the view of the proponents of the Copenhagen interpretation? Bohr and his followers had been avoiding the ultimate consequences of their claims for several years. In terms of locality, it had long been known that the processes around the collapse of the one-particle wave function (for example, when the electron hits the photographic plate) could only be explained by a non-local action. But nobody really wanted to tackle this hot issue! Nobody had explicitly asked whether the basic principle of locality was also supposed to apply in the quantum world.

Then came the double assault in 1935. First, Einstein and his two co-authors published the EPR paradox in which they expressed their opposition to the Copenhagen interpretation and thus got to the heart of the Bohr–Einstein debate. And then, encouraged by Einstein's thought experiment, Erwin Schrödinger, the other prominent critic of the Bohr–Heisenberg interpretation, brought the notion of entanglement into play.

Schrödinger's cat experiment essentially also deals with action at a distance: Only with the observation of the dead (living) cat does the nucleus "know" that it has (not) disintegrated. With this, Schrödinger translated the EPR paradox into the macroscopic world and thereby poured more oil onto the fire.

> Ten years after Schrödinger's equation and Heisenberg's matrix mechanics, the questions raised by Einstein, Podolsky, Rosen, and Schrödinger led physicists straight to the philosophical heart of the quantum world.

Bohr responded to the attack by Einstein, Podolsky, and Rosen immediately with an article of his own that bore exactly the same title: *Can Quantum-Mechanical Description of Physical Reality Be Considered Complete?* His answer was: Yes! Quantum mechanics was indeed complete and there were no hidden variables. He was supported by the great mathematician John von Neumann, who in 1932 had published a mathematical proof of the impossibility of a quantum theory with hidden variables. However, there were two issues with this proof. First, it was available only in German, so Neumann's work was known only to a minority of physicists. Second, it turned out that von Neumann had made some very specific assumptions and his proof was far from being universally valid.

Even though Bohr and his Copenhagen companions were ready to defend their interpretation of quantum physics to the last, Einstein's thoughts had stopped them in their tracks. If Bohr and his followers wanted to uphold the Copenhagen interpretation, they had to:

- either yield to Einstein's demand for hidden variables in the quantum world (and thus also to his reality hypothesis),
- or accept the existence of spooky actions at a distance.

This cleared the way for a deeper assessment of the concept of entanglement. The fact that this phenomenon had been given a specific name by Schrödinger was a first step. But how should physicists grasp this strange coupling of quantum particles (or cats and observers)? And how could a measurement have an instantaneous effect on the wave function of an entangled but distant particle?

> Ironically, the decisive impetus for the Copenhagen camp to finally realize the full meaning of the phenomenon of an instantaneous long-distance action came from its main opponents, Einstein and Schrödinger.

A Brief Flash in the Pan

However, the spirit of discovery was soon damped down. After the heated discussions of 1935, a silence arose regarding the question of which of the two scenarios was applicable:

1. Quantum physics is real, deterministic, and local. Even in the microcosm, states are objectively determined. The fact that experiments provide

a different picture is due to the fact that quantum theory is incomplete without the introduction of hidden variables (Einstein's position).

2. Quantum physics is non-real, non-deterministic, and non-local. The phenomenon of spooky long-distance actions really exists (position adopted by proponents of the Copenhagen interpretation).

Einstein and Schrödinger, but also their opponents, had all tried to leave the terrain of philosophical speculation and collect arguments from concrete (thought) experiments to support their respective positions. But they were forced to accept that they were unable to resolve the issues raised by the EPR paradox and Schrödinger's cat. The hidden variables, if they existed, would be hidden forever. After all, how could one assess something that was fundamentally unmeasurable or that itself changed under the measurement? Einstein, Schrödinger, and Bohr had to leave the problems of spooky actions at a distance, entangled particles, and half-dead cats open.

> The opposing camps could only discuss philosophical questions, just matters of taste, not hard experimental or mathematical facts.

Three questions thus remained unanswered:

1. Locality: Could quantum world systems only affect their immediate neighbourhood, or were there instantaneous effects at a distance?
2. Realism: Did quantum systems have any reality independently of the observer?
3. Hidden variables: Were there inaccessible variables that controlled the evolution of a quantum system and its wave function in a deterministic way?

The two sides were clear: on the one hand Einstein, who answered all these questions affirmatively, and on the other, the conventional quantum mechanics of the Copenhagen interpretation, represented by the vast majority of physicists, who answered these questions negatively. Although none of the contenders could claim a proof of their position (except for Neumann's mathematical proof, which turned out to be inadequate to the task), the subject would haunt the protagonists until the end of their lives (Einstein died in 1955, Schrödinger in 1961, Bohr in 1962). Even on the day of his death, Bohr had been working on possible gaps in the Copenhagen inter-pretation; the notes of his discussion with Einstein were found lying on his desk.

> The fathers of quantum physics never discovered the final answers to the questions posed by quantum theory. For half their working lives they had puzzled over them, without coming any closer to a solution.

The next generation of physicists, who were more concerned with physical phenomena and their mathematical description than with philosophical considerations, put the problem aside for a while. After all, everything was working fine with quantum physics:

- The theory of the electromagnetic field could be formulated as a quantum theory.
- Dirac's theory correctly predicted the existence of new particles.
- After decades of hard work, the abundance of new experimental results could be summarized in a consistent theory of all elementary particles, which involved beautiful mathematical structures and symmetries.

Faced with these great successes, physicists were not in the mood to deal with philosophical questions. They preferred to take care of the technological applications of what they had so far understood of quantum physics.

> For a long time—from the late 1930s to the 1960s—the EPR paradox and Schrödinger's cat were off the radar.

But then came a surprise. Based on work by the American physicist David Bohm, the Irish physicist John Bell succeeded in the 1960s in formulating an equation (or rather: an inequality), which potentially made it possible to determine *by experiment* whether the hidden variables predicted by Einstein existed or not. Bell's inequality became the prelude to a new and exciting second revolution in quantum physics, ignited by the phenomenon of entanglement.

24

The Experimental Resolution of the Bohr–Einstein Debate: How Entangled Particles Made Their Way from Theory into Practice

In the late 1940s and early 1950s, some American physicists came under the spotlight of the reactionary Senator McCarthy and became targets of his attacks, similar in some ways to those of the medieval Grand Inquisitors. The modern heresy was communism, and those who leaned towards liberal and leftist ideas were suspected of anti-American activities. Even Robert Oppenheimer, the father of the American atomic bomb, who had made a significant contribution to his country's becoming a world power, was persecuted by McCarthy and his followers.

The physics professor David Bohm also fell foul of the anti-communist witch hunts. In 1949, in order to protect his colleagues, he refused to testify before the "House Un-American Activities Committee". The then 32-year-old paid a high price for sticking to his principles. He was forced to leave the University of Princeton, where he had worked closely with Albert Einstein. Despite the intercession of his famous colleague, Bohm was expelled from the academic community of his home country. In 1951, in exile in Brazil, Bohm began to question the prevailing non-real and non-local interpretation of quantum physics. Since Einstein and Schrödinger, no one else had dared to do so.

> The physicist David Bohm was persecuted for political reasons in the United States. But even in exile, he was not intimidated by the prevailing opinions and developed a completely new interpretation of quantum mechanics.

© Springer Nature Switzerland AG 2018
L. Jaeger, *The Second Quantum Revolution*, https://doi.org/10.1007/978-3-319-98824-5_24

Bohm or Bohr?

Bohm had already proven that he was a remarkably independent mind in Princeton, where he and Einstein initially pursued the same goal. They wanted to show that quantum objects have a *reality independent of measurement*. They both opposed the generally accepted quantum theory, according to which there were no objective properties in the quantum world. Bohm also believed in Einstein's hidden variables, which would make events in the quantum world, not only real, but also deterministic. But physicists who supported the established view of a non-real and non-deterministic world would only shake their heads over such stubbornness.

However, Bohm and Bohr differed in one thing: Bohm's theory was *not local*. It did allow for instantaneous long-range effects, and in fact it even anticipated them. He thus adopted a completely new line. Mainstream physicists vehemently rejected the first two features of his real, deterministic, and non-local theory, and Einstein dismissed his colleague's theory as "too cheap" on the basis of the last.

> Bohm's interpretation of the quantum world was realistic, deterministic, and non-local—a combination that resonated neither with the many followers of the Copenhagen interpretation nor with Einstein.

It was especially interesting to see how Bohm explained that quantum particles could have real properties. He postulated that the wave properties observed in experiment are not a measurement-dependent expression of the electron, but represent actually existing waves in which equally real electrons, existing as particles, appear to "swim". These "guiding waves" also explained the instantaneous action at a distance. Because it was not possible to measure them *directly* as a physical quantity, Bohm initially described his theory as a hidden variable theory, whence his intellectual proximity with his colleague Albert Einstein.

To everyone's surprise, Bohm was able to show that the equations of his theory were able to perfectly predict and explain the results of all known experiments, just like the conventional, non-deterministic quantum theory of the Copenhagen interpretation. One could argue however one wanted about which of the two interpretations was the right one, but one thing could not be denied: there was definitely more than *one* theory that could explain the phenomena of the quantum world.

Bohm's theory of nonlocal quantum physics rivalled the generally accepted Copenhagen interpretation. Both interpretations were consistent with experimental observations in the micro world.

The Wheel Invented Twice

Unfortunately, Bohm was not an avid reader of other scientists' publications, otherwise he would have known that he had basically reformulated an interpretation by the French theorist Louis de Broglie from 1925. De Broglie, too, had interpreted the observed wave character of electrons as physically real; he had called those guiding waves "pilot waves" (*ondes pilotes*). But Wolfgang Pauli, one of the leading minds in quantum physics, had vehemently rejected this theory and convinced de Broglie at the time to discard it. In order to do justice to de Broglie's idea, Bohm's renewed interpretation of the wave function is called the *de Broglie–Bohm Theory* today.

Not surprisingly, Bohm's new theory met with scepticism and even fierce opposition from large parts of the physics community. Heisenberg called the pilot or guiding waves a "superfluous ideological superstructure". With his usual candour, Pauli, who had already talked de Broglie out of the idea of pilot waves 25 years earlier, spoke of "artificial metaphysics". In one respect, however, the criticism of the new theory was not entirely unjustified: it did not predict a single new phenomenon and introduced a whole set of unverifiable components. In addition, the pilot wave for systems with more than one particle would move in higher-dimensional spaces and thus lose all intuitive descriptiveness (see Chap. 9).[1] This is less problematic in an abstract interpretation of the wave function as a probability wave than for its concrete interpretation as a physical entity as in Bohm's theory.

The fact that a new theory is put to test by other physicists, or even attacked by them, is quite normal and also necessary to separate the wheat from the chaff, as in all sciences. The worst thing for Bohm was that, after the first excitement, hardly anyone took any deeper interest in his new interpretation. The scientific guild, pragmatically shaped by a new generation of physicists, preferred to study the much more exciting new fields of quantum field theories and elementary particle physics. There was no career to be had with questions about the basics of quantum theory. For most physicists, that

---

[1]That is exactly what characterizes entanglement, and what constitutes the difference with classical physics. There, the state spaces of many-body systems could always be separated into the subspaces of the individual particles. This is what is no longer possible in the quantum world.

would just be a topic for philosophical small talk, and they disparagingly referred to Bohm's field of research as *armchair philosophy.*

Bohm had proved that quantum mechanics did not necessarily have to be interpreted in the conventional way, but his work did not receive wide attention.

From the Idea to a Concrete Plan

In 1957, Bohm made another crucial contribution to the question of whether the apparent action at a distance between entangled particles really corresponds to a non-real nature of the micro world or whether it is not mediated by hidden variables. Together with his student Yakir Aharonov, he reworded the EPR thought experiment. Instead of the position–momentum states of any unspecified quantum particles, as described by the original EPR paradox, he proposed electrons and their spins as protagonists. However, this proposal initially did not stir up much interest, either.

Today we know that Bohm's variation of the EPR thought experiment was a significant step on the path from pure theory to concrete physics.

Bohm himself wrote:

> *This experiment* [author's note: entangled spins] *could be considered the first clear empirical proof that the aspects of quantum theory discussed by Einstein, Rosen, and Podolsky represent real properties of matter.*[2]

In fact, many years later, it became possible to study two entangled electrons in a real experiment in the manner suggested by Bohm. Bohm was 65 years old when he experienced this triumph.

This is how the modified EPR experiment works. Two electrons are entangled. They are fermions so they obey the Pauli principle, i.e., they have opposite spins:

---

[2]D. Bohm, Y. Aharonov, Discussion of Experimental Proof for the Paradox of Einstein, Rosen, and Podolsky, *Physical Review*, 108, 1070 (1957).

- electron 1 with spin *up*, electron 2 with spin *down*, or
- electron 1 with spin *down*, electron 2 with spin *up*.

Before the measurement, it is unknown which electron has which spin. Their common state is a superposition of the two possible states and is described by a *single* wave function, that is, they are entangled. It is only known that their total spin is zero, being made up of $+\frac{1}{2}$ for one electron and $-\frac{1}{2}$ for the other.

Now the entangled electrons move away from each other, one being brought to location A, the other to location B. A and B can be any distance apart. Only now the spin of one of the two particles is determined, for example, the one at location A. We then have the following situation:

- Upon measurement, the wave function, previously a superposition of the *up* and *down* spin states of the two particles, collapses.
- For example, the measurement determines that the electron at point A has spin up.
- Without delay, the wave function of the particle at location B also collapses, and because the first electron was "caught" in the *up* state, the second must show the state *down*.

Although Bohm's experiment could not yet determine the nature of the quantum world is (local or non-local, real or non-real), his idea stimulated the legitimate hope that physicists would someday be able to carry out a real experiment on the properties of entanglement.

Turn Two into Four

It is 1964. Everyone working in quantum physics is under the sway of the Copenhagen interpretation. Everyone? No. Some indomitable physicists will not give up their attempts to improve on Bohr's solution. But the vast majority of them are holding their positions in the strongholds at CERN, CalTech, MIT, and Stanford, holding any rebels at bay.

In this rather unequal dispute, two positions meet, which the reader already knows from the last chapter:

1. The quantum world is *not realistic*, since the properties of quantum objects depend on their measurement. And it is also not local, because there is an instantaneous effect acting on widely separated entangled particles. This interpretation is the predominant one.

2. The quantum world is realistic and local, quantum objects have real properties that are independent of any measurements that may take place. Because in a local world spatially separated events cannot influence each other, hidden variables are held responsible for the observed long-distance effect. This preserves the principle that effects always take place with some time delay after the causes.

In the 1950s, Bohm's attempt (realistic, non-local) to escape from this scheme was fought off by both sides. But at this point the Northern Irish physicist John Bell came on the scene. He was the first to understand and accept the meaning of Bohm's work in all its detail.

> Bell, like Bohm, recognized that there were more ways of interpreting quantum mechanics than the non-realistic and non-local Copenhagen interpretation and its realistic and local opposite.

Theoretically, with the attributes local/nonlocal and realistic/non-realistic, one can construe *four* possible combinations:

1. Quantum mechanics is a *local* and *non-realistic* theory. This option was initially Bohr's favourite and was thus the original Copenhagen interpretation. However, Bohr was unable to explain how and why wave functions collapse and said that at sufficiently large separations two entangled particles would somehow decay into their individual wave functions. By the thought experiment of Einstein, Podolsky, and Rosen (the EPR paradox) of 1935, he had been forced to abandon locality and advocate a second possible combination.
2. Quantum mechanics is a *nonlocal* and *non-realistic* theory. With a few exceptions, this had been the accepted reading of quantum physics until the second half of the 20th century—as the name "standard quantum mechanics" illustrates.
3. Quantum mechanics is a *local* and *realistic* theory. Only a few physicists upheld this possibility, including Einstein.
4. Quantum mechanics is a *nonlocal* and *realistic* theory. This was the perspective adopted by David Bohm with his pilot wave theory. Although it was able to explain all previously known properties of the quantum world just as well as standard quantum mechanics, this possibility was widely ignored.

Two parameters yield four possibilities—what today seems so simple and obvious was at that time hard to introduce into the discussion. The possibility that the world might be real as in cases 3 and 4 above was ruled out at the outset. Mainstream physicists formed a closed community. Anyone who broke its unwritten rules (and was not called Einstein) would have problems getting financial support for research or even for a teaching job. But why did the physics community insist that a real quantum world was impossible? Because that would automatically mean the existence of hidden variables. And because such hidden parameters would live up to their name and would not be measurable, in principle, most physicists would apply Ockham's razor,[3] saying that it should not be necessary to deal with them.

> The idea that the quantum world could be realistic was largely neglected by the physics community around the middle of the 20th century.

Bell was a follower of Bohm's and de Broglie's pilot wave theory. He wrote:

> This idea seems to me so natural and simple, to resolve the wave–particle dilemma in such a clear and ordinary way, that it is a great mystery to me that it was so generally ignored.[4]

But Bell did not want to theorize and speculate like others, because that had led nowhere in the past decades. He was looking for a way to find clear experimental evidence for one or other of the positions.

A Successful Sabbatical Year

Since 1960, Bell had been at CERN in Geneva working on questions of elementary particle physics and quantum field theory, i.e., on exactly the hotspot topics in physics at the time. He made significant contributions in these fields, which were followed up by many of his colleagues—some of whom subsequently earned Nobel Prizes. Bell's secret passion since his student days, however, had been the interpretation of quantum mechanics.

---

[3]According to this principle, between two competing theories which both describe observations equally well, one should choose the one that is simpler or comes with fewer assumptions.

[4]J. Bell, *Speakable and Unspeakable in Quantum Mechanics*, Cambridge (1987), S. 191.

In 1963, Bell took a one-year sabbatical from CERN. He took advantage of his freedom and went to the US, and to Stanford among other places. Here he was finally able to concentrate on the fundamental questions of quantum theory. He knew that Bohm had revealed gaps in the standard interpretation. These he wanted to close.

Bell was aware of the danger that his work on the fundamental questions of quantum physics would be considered exotic by the physics community, or that he would even be seen as a troublemaker and lose his reputation as a scientist.

Bell began by questioning von Neumann's mathematical proof of the impossibility of hidden variables in any possible quantum theory, which he had given in 1932 as a 29-year-old whiz kid mathematician (but which had not stopped Einstein from persevering in his claims about their existence). Bell showed that, although Neumann's argument was mathematically correct, it was based on unrealistic physical assumptions. From today's perspective, it is hard to understand how physicists could have followed von Neumann's proof so blindly for more than thirty years.[5]

His result allowed Bell to search empirically for the hidden variables. It is not uncommon in science that those who try hardest to refute a theory should ultimately help to strengthen it. Bell had the same experience: his work ultimately led to hidden variables (at least *local* ones) finally being buried.

Bell took up the EPR line of thought in David Bohm's version, which is why he entitled his paper *On the Einstein–Podolsky–Rosen Paradox*. And he finally succeeded in formulating the crucial mathematical relationship that would make the question of the nature of the quantum world experimentally accessible. The following section deals more closely with the logical foundations of his thoughts. The reader can choose to skip this since it is not essential for understanding what follows. However, those who are willing to follow Bell's train of thought will experience a fine example of the beauty and elegance of ground-breaking ideas in physics.

---

[5]In 1935, the mathematician (and student of Emmy Noether) Grete Hermann had already recognized the problems in von Neumann's argumentation. However, her work was only rediscovered in 1974.

There Really Are Spooks in the Quantum World

The following were Bell's considerations regarding Bohm's EPR thought experiment:

A particle with zero spin decays into two entangled particles with spins $+\frac{1}{2}$ and $-\frac{1}{2}$. Their spins can be illustrated by certain directions of rotation, for example "up" and "down". In this case the axis of rotation is vertical. But there are also other possible axes, for example, a horizontal axis, which makes the spins rotate "left" or "right". In fact, there are infinitely many axes around which a particle can spin, and its spin be measured. However, along any observed axis, the quantum laws ensure that the spin can only take the value $+\frac{1}{2}$ or $-\frac{1}{2}$.

In the experiment, the observer can freely choose the axes used to measure the spins of the two electrons. When the spin is measured along a certain axis, its value changes from a statistical probability to a definite property.

> Before the measurement, the spin of either electron in any given direction is indefinite. Only probabilities apply. However, the theory ensures that these entangled electrons always have complementary spins along the same axes of rotation.

Of course, we never measure just one electron pair in an experiment. And we don't just measure along *one* axis, but along two axes, one for each of the two entangled particles. Then depending on the setting of the two measurement directions, we obtain a certain distribution of the measured spin values. Regardless of which two axes the experimenter decides to use, the following statements apply:

- Due to the entanglement, for each particle measured with spin $+\frac{1}{2}$ along a given axis, the partner particle will have spin $-\frac{1}{2}$ along that axis, and vice versa.
- If the spins of many entangled particles are measured along different axes **a** and **b** (bold type indicates that this is a direction vector), the results of the measurements are correlated. The closer the chosen directions of the axes **a** and **b**, the higher the correlation of the measured spin values (without entanglement, the measurement results for the spins would be completely independent of one another, i.e., the measured values would be completely uncorrelated).

- However, this measured correlation does not yet say much about where it comes from. For both alternatives (instantaneous action at a distance or hidden variables), the measurement results are the same.

> A non-local theory and a local theory with hidden variables cannot be distinguished by measurements along two axes.

At this point Bell came up with a brilliant idea. He considered *three axes*, i.e., the spin values as measured in arbitrary directions **a**, **b**, and **c** (where **a**, **b**, and **c** must not be coplanar). Specifically, he considered the probabilities with which the measured spins of the entangled particles take on specific values simultaneously along two of the three axes. Here are two examples:

- The result P(**a**＝+½, **b**＝+½) represents a spin of +½ in direction **a** and +½ in direction **b**. P stands for "probability".
- P(**b**＝−½, **c**＝ +½) describes the measurement of a spin −½ in direction **b** and +½ in direction **c**.

Analogous to the measurements along two axes, the measurement on a particle, for example in direction **a**, determines the spin of its partner particle along *the same* axis exactly, but not along the measuring directions **b** or **c**. In this case again only statistical correlations apply. But here a different picture suddenly emerges compared to the case of only two measurement directions: the relationship between P(**a**, **b**), P(**a**, **c**), and P(**b**, **c**) in a local theory with hidden variables differs from the one in standard quantum theory! Bell succeeded in establishing the crucial relationship. For a local theory with hidden variables, one must always have:

$$P(\mathbf{a},\mathbf{b}) \leq P(\mathbf{a},\mathbf{c}) + P(\mathbf{b},\neg\mathbf{c})$$

where P(**b**, ¬**c**) denotes the probability that the spin is measured in the **b** direction, but not in the **c** direction with the respective values of +½ or −½.

> With Bell's inequality, the local realistic version of quantum theory with hidden variables can be *measurably* differentiated from the standard non-local quantum theory.

Bell's inequality is a no-go theorem: if the underlying theory (existence of hidden variables) turns out to be true, experiments must not yield results that expose the mathematical relationship as false. In the case of Bell's inequality this means:

- One first assumes that, before any measurement, the reality principle and hidden variables determine in which directions the spins of the electrons are measured. For this assumption to be true, the inequality *must* hold.[6]
- Only if there are no hidden variables would it be possible for the inequality to be violated.[7]

> Bell's inequality had enormous significance: it brought the question of the nature of the quantum world from the realm of theory to within reach of experiment. Thought experiments thus became experimentally verifiable statements.

Now it was turn of the experimentalists. Bell's inequality was an appeal to physicists to solve the open questions by means of experiment.

A Starting Signal—But Hardly Anyone Moves

Bell had set a new gold standard in the discussion of the nature of the quantum world: experiment. But he had to endure the same cool response as Bohm had done ten years earlier: the scientific community reacted to his inequality in a very measured way. Because hardly anyone was interested in the basics of quantum theory any longer, he could only publish his work in second-rate journals. Accordingly, his ideas were not widely noticed. In addition, after his sabbatical, Bell went back to CERN, where he returned to his official duties. It was not until five years later, around 1970, that he published again on the subject that was so close to his heart. And again he had to publish in a journal that led a rather underground existence in the physics community.

---

[6]Bell's inequality only applies to *local* hidden variables. Nonlocal hidden variables are still possible if the inequality does not hold. The best known theory with non-local variables is the aforementioned de Broglie–Bohm theory.

[7]More specifically, Bell's inequality relates only to *local* hidden variables. Non-local hidden variables are still possible with the inequality being violated. The best known theory with non-local variables is the aforementioned de Broglie–Bohm theory.

It was initially only a small group of physicists who became aware of Bell's work and began to consider the basic questions of quantum physics, but their endeavours were far from being accepted by the broader physics community.

The field explored by the "Bellians" is best described as "experimental philosophy", because *actual* experiments on Bell's inequality were still out of reach at the time. First, the vibrant little group had to learn how to handle entangled particles. Only in the early 1980s were the first teams ready to carry out experiments that allowed reliable conclusions about the nature of the quantum world. And even then, it would be some time before these results were finally noticed by mainstream physics. Even in the 1990s—the present author (engaged at the time in the very hip chaos theory) can confirm this from his own observation—quantum theorists had to put up with statements like: "Does what you work on make any sense? Use your time for something more useful!"

With Light to Success

There were several approaches to the study of Bell's inequality. But when transposed to a suitable concrete experimental setup, most of them proved to be rather unwieldy. Two prominent examples were:

- As suggested by Bell (and Bohm), one of the setups involved entangling two electrons with spin $+\frac{1}{2}$ or $-\frac{1}{2}$, respectively.
- Another possibility was to entangle the spins of atomic nuclei. If a diatomic molecule (for example, hydrogen H_2, helium He_2, lithium Li_2, or chlorine Cl_2) of spin zero is excited with a laser of sufficient energy, it will break apart. The nuclear spins of the atoms thereby created are entangled.

But there was another setup that was experimentally much easier to handle than those involving electrons or atomic nuclei and their spins. These used photons and the direction in which they vibrate, their so-called polarization.

Electromagnetic waves, for example light waves, consist of oscillating electric and magnetic fields. Their vibrations occur in a plane which is always perpendicular to the direction of propagation of the wave. Individual photons can also be assigned such an orientation. For example, they may be polarized vertically (90°) or horizontally (0°), but also at any other angle.

As for entangled electrons whose spin properties are related, photons can be entangled via their polarization. There is one difference, however: while two entangled electrons can never have the same spins due to the Pauli principle, the polarization directions of entangled photons are identical.

The first entangled photon pairs were realized in 1967 by Carl Kocher and Eugene Commins of the University of California at Berkeley. It turned out that they were comparatively easy to deal with:

- They are easy to produce. A laser beam is used to shoot a photon with high energy onto a so-called nonlinear optical crystal. From the high energy photon thus emerges a pair of two photons, each with half the energy of the original photon.
- Their polarizations are entangled as desired.
- Due to the conservation of momentum they are emitted in opposite directions and can thus be easily detected.
- They are easy to transport in fibre optic cables.
- Their polarizations are easy to measure and manipulate with polarizing filters.

With such entangled photons, Bell's inequality was experimentally much easier to examine than with electrons. The probabilities P(a, b), P(a, c), and P(b, ¬c) of Bell's inequality thereby translate into count rates N(a, b), N(a, c), and N(b, ¬c).

> The use of entangled photons instead of electrons was a breakthrough in trying to prove the validity of Bell's inequality by experiment.

The Gradual Demise of Hidden Variables

Over time, the experiments devised by the small group of pioneers to explore the nature of the quantum world became more and more sophisticated. However great the antagonism between the elegance of theoretical physics and messing around with tiny calibrations for experimental setups in stuffy, windowless laboratories may appear to be, the ingenuity and endurance of the experimenters who worked on Bell's inequality to provide an empirical test of the thesis of a local quantum theory, lived up to that of any theoretical physicist.

> The path to experimental verification of Bell's inequality is a prime example of the creativity and inventiveness of experimental physics.

To Bell's surprise, more and more evidence emerged that his inequality was indeed violated! Hidden variables were losing the game. But doubt remained, because, despite the elaborate experimental designs, there were still inaccuracies and thus theoretical loopholes whereby hidden variables might just be allowed in through the back door, as it were.[8] The path towards a clear-cut proof was still long. Almost two decades of tireless work were undertaken between the emergence of Bell's inequality and its unequivocal experimental verification.

> It was not until 1982 that the French physicist Alain Aspect and his team proved beyond any doubt by a concrete experiment that Bell's inequality was being violated.

Rather like David Bohm and John Bell, Alain Aspect had also found his interest in the foundations of quantum mechanics during an extended stay abroad. In his case, it was a teaching assignment in Cameroon from 1971 to 1974. Aspect's idea was in principle straightforward, but experimentally it was anything but easy to implement. Previously, the orientations of the polarizing filters had always remained the same during the experiments. Aspect now changed their direction *during the flight of the photons*. This made it impossible for any information to be exchanged between the two measuring devices *before* the measurement.

With his brilliant experimental setup, Aspect succeeded in perfectly validating the violation of Bell's inequality, and the last doubts about the non-locality of the quantum world were thus dispelled. Entanglement and hence also nonlocality is an incontestable, integral part of the quantum world, even over great distances. This was now attested, not by "mere" philosophical arguments, but rather by hard mathematical and empirical facts.

---

[8]For example, in 1972 an experiment with entangled photons by Stuart Freedman and John Clauser showed a violation of an inequality that was very similar to Bell's inequality, but still contained some loopholes for a local quantum theory with hidden variables. Nevertheless, this so-called CHSH (Clauser–Horn–Shimony–Holt) inference provided some first experimental evidence for the non-locality of the quantum world.

The big quarrel between Einstein and Bohr, whether or not there are spooky actions at a distance in the quantum world, was ultimately decided in favour of Bohr: they do exist. The idea of hidden variables had to be buried.

A World of No-Gos

Continuing right up until the present day, experimental physicists have come up with more and more refined experiments aiming at uncovering more and more of the strange properties of quantum world. After realizing that it was not necessarily local, physicists wanted to answer the other controversial question of quantum theory once and for all: they wanted to determine *by experiment* whether it is real or non-real, that is, whether properties of quantum entities are independent of any observation. Over time, other no-go theorems emerged, and so far the corresponding experiments have shown that it is highly unlikely that the quantum world is real. Even so, some physicists and philosophers have not given up on the idea that quantum particles have an independent reality. They ask themselves: "Could there possibly still be a connection that we have overlooked?"

Once physics had finally been forced to say goodbye to the idea of a local quantum world, the possibilities of "realistic" interpretations are becoming increasingly limited today.

There is much to suggest that the quantum world is a world without an independent existence of things. It is rather an objectively indeterminate world full of incomplete information, determined only by its interaction with the environment. But the "Einsteinians" continue to insist with astounding resistance on the existence of something essentially real in the quantum world. On the other hand, this would have to be a very strange reality that has little in common with the reality we experience every day. Because whoever wants to save realism must be prepared to accept some very strange features. For example, it has been shown that, in a real quantum world, quantum influences must travel faster than light, or even backwards in time.

However we consider it, either we abandon the idea of an independently real quantum world, or we accept that this reality must be completely different from anything we can imagine.

25

The Age of Entanglement: From Spooks to a New Quantum Revolution

As soon as the generation, manipulation, and measurement of entangled quantum particles had become a day-to-day business in basic research and it became apparent that this knowledge could also be used for completely new technologies, it was like a dam breaking. Within just a few years, the theoretical framework for a multitude of exciting new technologies developed out of the exotic debate about the nature of the quantum world that for so long had not been taken seriously by mainstream physics. In the late 1980s and 1990s, a new generation of quantum physicists succeeded in developing the first concrete applications of entangled quantum particles with ever more sophisticated experiments and measurements. Their final breakthrough, however, was to take another 20 years.

> From the moment Bell's followers learned how to deal with entangled particles, the tide gradually turned: the ugly duckling turned into a beautiful swan in the form of one of today's most exciting fields of future technology.

Quite suddenly, those who had previously been ridiculed could even make a career outside of basic academic research with their expertise. One of them is the Swiss physicist Nicolas Gisin. In 1984 Gisin moved from university to a start-up producing fibre optic technology for telecom companies. Later he became a professor in Geneva, but has remained very close to industrial research ever since. In 2001 Gisin founded *idQuantique*, today one of the leading companies in the field of quantum information and

© Springer Nature Switzerland AG 2018
L. Jaeger, *The Second Quantum Revolution*, https://doi.org/10.1007/978-3-319-98824-5_25

communication. Gisin holds numerous patents on technological applications of entanglement-based quantum-mechanical technologies.[1]

- In 1997, Gisin and his group were able to demonstrate quantum nonlocality for the first time beyond strictly controlled laboratory conditions. Supported by the Swiss telephone company Swisscom and using standard fibre optic cables, they managed to separate entangled pairs of particles over a distance of more than 10 km, in fact, on either side of Lake Geneva, and measure their properties.
- In the early 2000s, Gisin succeeded in teleporting quantum states, i.e., transmitting qubits, over even longer distances.

Another pioneer of quantum teleportation is the Austrian quantum physicist Anton Zeilinger, whose experiments with entangled photons over long distances have earned him the nickname "Mr. Beam". In 1997 Zeilinger and his group succeeded in performing the first demonstration of quantum teleportation of the state of an independent photon.[2]

> Thanks to physicists like Gisin and Zeilinger, it is today possible to create and measure entangled states over distances of hundreds of kilometres.

First Steps Towards a New Technology

What was Gisin's intention with these experiments? The experimental confirmation that quantum theory allows instantaneous remote effects suggested that, beyond classical information processing, superpositions and quantum-mechanical entanglement might be used to efficiently and quickly transfer and process information about individual quantum states. This idea was the birth of quantum information technology. It deals in qubits, superpositions of binary states (see Chap. 4), rather than classical bits (information packed in zeros and ones) for information processing.

In a first wave of exuberance, some quantum technologists even believed that they could transmit classical information instantaneously and thus

---

[1]See also Gisin's book N. Gisin, *Quantum Chance—Nonlocality, Teleportation and other Quantum Marvels*, Springer, New York, Heidelberg (2012).

[2]D. Bouwmeester, J. W. Pan, K. Mattle, M. Eibl, H. Weinfurter & A. Zeilinger, Experimental *Quantum Teleportation*, *Nature*, 390, 575–579 (1997).

faster than light, which would have invalidated a basic principle of special relativity. Even the abolition of causality was suddenly conceivable, in the sense that an effect might appear earlier than its cause. And even time travel seemed possible in principle.

> Instantaneous communication, abolition of causality, time travel... all these possibilities that Einstein had fought against during his lifetime seemed suddenly to be within reach in the quantum world.

However, it soon became apparent that these ideas would have to remain in the field of science fiction. The reason for this is the so-called *no-signalling theorem*. This states that measurements on quantum mechanical systems cannot be used to transmit definite (classical) information from one observer to another without a time delay. Three features ensure that the theory of relativity and causality are also preserved in the quantum world:

- Objectively indeterminate states of a particle can never be carriers of definite information. For the output of the measurement on the first of the entangled particles completely depends on chance, whence also the state of the second particle.
- Once a measurement is performed on a particle, the entangled particle is instantaneously in the same (photons) or in the opposite (electrons) state,[3] but the *information* about the actual state of the two particles still has to be transmitted to the observer. The reading, recording, transmission, etc., of the information can only be done in the conventional classical way. After the switch from indeterminate quantum information to determined classical information, the classical rules of time and causality apply again.
- It is not possible to transmit classical bits via entanglement from the transmitter to the receiver particle. In theory, that would not be difficult: the sender could encode his message by taking a measurement whenever he wants to send a 1. If the receiver is supposed to receive a 0, the sender performs no measurement. At first the receiver does not know whether the transmitter measured his particle (i.e., sent a 1), but he can find out by copying the state of his own particle several times and then measuring the

---

[3]There are also entangled states in which electrons have the same spin and photons different polarizations. A rare case is the parallel alignment of the individual spins in Cooper pair electrons, in which their total spin thus sums to one. Here, physicists speak of a triplet state.

copies one by one. If all copies are in the same state, the receiver knows that the transmitter previously sent a 1 (since with the measurement on the sender's particle, the receiver particle will also have entered a definite state). If the sender did not measure his particle, the measured states of the receiver's particle copies will be statistically distributed equally.

> A prerequisite for the instantaneous transmission of bits would be that the entangled quantum states could be copied. However, the so-called *no-cloning theorem* prohibits just that.

The no-cloning theorem follows directly from the absence of an independent reality in the quantum world. When measuring a quantum system in a previously indeterminate state, it is put into a new (determined) state. But then it is no longer the same (indefinite) particle as before. Thus it cannot be copied. Because a statistical check cannot take place, no instantaneous transmission of bits is possible.

Of course, the fact that quantum particles cannot be copied applies only to unknown states. If we already know that a particle is in a certain state, we can easily make copies of it. But this case is uninteresting for the transmission of information.

> Despite entanglement in the quantum world, no information can be transmitted faster than light. The order of events and thus the basic rules of causality are preserved.

Is that the end of quantum information technology? In fact, far from it. For even if an instantaneous transfer of information is not possible, there are great advantages in using entangled states in combination with a classical information channel. An example is their use for absolutely tap-proof information transmission.

Tap-Proof Transmission Through Space

In almost every aspect of our everyday lives, data security plays a fundamental role, from fighting terrorism and securing energy supplies to protecting the data concerning our private bank accounts or on our cell phones. So far, security of encryption has always been based on maintaining the secrecy of a key:

- The ancient Greeks used the so-called *Skytale* method to encrypt secret messages. For this purpose, they wrote the text across a tape wound around a wooden staff. Only when the recipient wrapped the ribbon around a wooden stick of exactly the same thickness could the text be read.
- Julius Caesar chose the method of character shift. The key provided the information about how many letters in the alphabet the recipient had to go forwards or backwards in order to decipher each letter.
- The outcome of the Second World War was decided in part by the decoding of the hitherto most complicated cipher machine, known as *Enigma*, which was used by the German military intelligence.

But *every* key can be cracked, even the most complex algorithms are not safe from unauthorized access, since they are all based on a deterministic algorithm. That's why the new cryptology technologies take encryption to a whole new level. In classical information transmission, a spy can listen in without the sender and receiver noticing it. But this is different with quantum cryptography. Instead of classical bits, qubits, i.e., superpositions of states between 0 and 1, are used. According to the no-cloning theorem these cannot be measured without being changed. Therefore any observation of them leaves traces. Both the receiver and the sender know that they are being intercepted and can change the key immediately.

In most cases, photons are used to produce entangled particles. These can easily be transported by optical fibre cables over long distances and are effortlessly measurable by means of polarizing filters. In Switzerland, quantum cryptography was first used in 2007 in the elections for the National Council in the canton of Geneva to secure the counting of votes against unauthorized interference. The person responsible for this was the above-mentioned quantum computer scientist Nicolas Gisin.

Transmitters and receivers, of course, do not only require security against unwanted listeners for the transmission of "normal" information. The Achilles heel in cryptology is the transmission of the key. When the key is intercepted, the most sophisticated algorithms become useless. However, if entangled particles are used to transmit the key, it cannot fall into the wrong hands without the sender and receiver noticing. In addition to its integrity, entanglement also guarantees the functionality of the key: because there are no transmission errors with entangled states, sender and receiver *always* have a matching key. Even if the measured state of a particle is completely random, the measurement on the other particle always gives the same (or exactly complementary) result.[4]

> With the aid of quantum information technology, every instance of eavesdropping and every attempt to pick up the key is automatically detected. Transmitters and receivers can always stay one step ahead of the spy by immediately changing the code.

However, the big breakthrough in quantum cryptography still lies ahead. Losses in fibre optic cables are still too high over long distances and the entanglement dissolves too quickly for a generalized usage of quantum cryptology technology. But there is another medium apart from glass fibre that does not induce as much interference: outer space. Since 2016, a satellite has been circling the earth, creating entangled pairs of photons, sending one of them to a receiving station on Earth, and thereby creating a secure connection between heaven and Earth. In 2017, the Chinese research group headed by Jian-Wei Pan, a student of Anton Zeilinger, succeeded for the first time in teleporting photons between a base station on the ground and a satellite in orbit.

From Classical Information Theory to the Quantum Computer

Quantum effects help to secure information transfer. But researchers are also working on ways to use them for the opposite: cracking conventional encryption.

The encryption methods used today, which have so far been considered safe, are based on prime numbers. The principle is quite simple: sender and receiver know two primes; the product of these primes is the number n. Using n, the message is encrypted in a certain way that can only be decoded by knowing the prime factors of n.

---

[4]However, entanglement is inherently rather unstable, i.e., sensitive to external disturbances. Therefore, in practice this statement is not always correct. Quantum information technologies require appropriate correction algorithms.

> In today's cryptology numbers with a few hundred digits are used. Factorizing them into their two prime factors would take a thousand years, even with the most powerful of today's computers.

This is because there is no algorithm yet to find the primes, except to try them all out, and conventional computers have to work through that list sequentially, bit by bit. With a quantum computers, however, things would look very different. For they are not just about an upgrade of conventional computers, but are based on a completely new information technology. One potential advantage is that, when trying out all primes to see if they are factors, they can work in a highly parallel manner. Let us take a closer look at the two different kinds of computer (see also Chap. 4).

Conventional computers use bits as the smallest possible unit of information. These represent a choice between two equally probable options (0 or 1). The von Neumann architecture described in Chap. 4 carries out the computation steps sequentially, i.e., bit by bit. Although today's computers contain components that are so small that quantum effects play a major role in them, this does not change the fact that their *functionality* is based entirely on the principles of classical physics.

> Today's computers are machines which, in principle, obey the rules of classical physics. Moreover, the information theory based on the fundamental unit bit can be described as classical.

In contrast, *quantum computers* are subject to a completely different information theory. Their basic unit of information is, as we saw, the qubit. The secret of the incredible speed of quantum computers lies in the entanglement of several qubits, which enables high-level parallel processing. While the conventional bit can assume the two states 0 or 1, a qubit consists of the information set which represents any superposition of the two states (see also Chap. 4). Thus, it can assume an infinite number of different states at the same time, and upon measurement, each of these would be realized with a certain probability, while prior to measurement they could all be processed simultaneously with a suitable algorithm. With such parallel computing power, for example, testing for prime factors works much faster. The two basic units—bits and qubits—are so different from each other that a qubit cannot be represented by conventional bits.

Another difference is that, in contrast to what happens with conventional computers, when a quantum computer reads out a result, this destroys the previously created coherent state containing all the information. One cannot store any intermediate results, because each reading of the information in the quantum states cancels the calculation so far.

A quantum computer performs its calculations according to a completely different logic than a conventional computer. Since the speed in factorizing large numbers as a product of primes is several orders of magnitude faster than that of any conventional computer, any standard numerical key becomes vulnerable.

A Missed Opportunity

It may seem strange that quantum computers were not built a long time ago. After all, quantum theory was well established long before the invention of modern computers. Nevertheless, decades passed by before physicists embraced the possibilities of quantum information processing. One reason is clear: for a long time, neither physicists nor computer scientists had the technology to deal with the phenomena of superposition and entanglement. And even today it remains very difficult to control them.

But there is a second reason. In the 1940s, the American mathematician Claude Shannon founded classical information theory, which is based on the processing of bits. His essay *A Mathematical Theory of Communication* is still considered the bible of the information age, and is one of the most influential scholarly works of the 20th century. The problem for quantum computers was that, for a long time, computer scientists took it for granted that Shannon's principle of bits must apply to *every* form of information processing.

Only later did computer scientists realize that there exist information concepts that go beyond bits and classical physics, and that the processing of qubits requires a completely new theoretical foundation.

Quantum information theory deals explicitly with the superposition and entanglement of quantum states. These properties have no equivalent in classical information theory, and a suitable new information theory was not created until the late 1990s, through the joint efforts of physicists and information theorists. These include David Deutsch (Deutsch Algorithm, 1985),

Peter Shor (Shor Algorithm, 1994), and Lov Grover (Grover Algorithm, 1996). Their work constitutes the theoretical foundation for possible future quantum computers.

It is also interesting to see how classical data processing and the concept of quantum computers might be connected. It seems likely that they will not stand on an equal footing. It is well known that quantum theory is the more fundamental theory, and indeed classical physics can be derived from it. Hence, quantum computer scientists expect the properties of conventional computers to turn out to be a derivative of the more fundamental properties of quantum computers, and Shannon's classical information theory to be derivable from quantum information theory.

> Quantum computers have their own rules for processing information. The functioning of the classical binary computer could turn out to be based on a "slimmed down" version of a superior quantum information theory.

A Wide Field

As lasers, transistors, and the resulting technologies of electronics, data processing, and communication show, quantum technologies have already become indispensable in our everyday lives. They all have one property in common: they are based on specific quantum properties of *independent* (i.e., non-entangled) particles in large ensembles. They employ the statistical properties and effects of many-body quantum systems. These include, for example:

- The tunnel effect in modern transistors (including floating-gate transistors). The most important technical device of our time, the computer, could not operate without this quantum effect.
- Coherence of photons in lasers.
- Spin properties in magnetic resonance imaging.
- Discrete quantum leaps in the atomic clock.

Now another quantum feature has entered the game, and before long it will very likely overshadow all the others: entanglement. Since physicists have been able to actively prepare, transfer, and process individual particles in entangled states, quantum technology is about to make a giant second leap. Here are some applications of entanglement, several of which have already been discussed in the first part of the book.

- Highly sensitive quantum sensors. The extreme sensitivity of entanglement to external influences allows for much more accurate measurement of time, gravitational forces, and electromagnetic fields. It will be possible to make clocks with even smaller error than today's atomic clocks, to detect mineral resources in seams deep below the surface of the Earth with great reliability, and to monitor cell activity in biological systems, for example, brain waves, with unprecedented accuracy.
- Producing new biocatalysts and drugs based on highly accurate simulations of complex chemical and biochemical processes.
- Replication of (quantum) biological systems and processes, such as in the production of an artificial leaf for energy conversion by photosynthesis.
- Secure communication through quantum cryptography.
- An absolutely random number generator: The fundamental uncertainty in the quantum world guarantees absolute randomness in the measurement of a quantum system. Classically generated random numbers, on the other hand, are actually pseudorandom numbers, because they are produced using well-defined deterministic algorithms.
- A new era of calculation and data processing with the already often mentioned development of the quantum computer.
- Quantum information transfer. This includes the ability to transport quantum information (qubits) over large spatial distances, often referred to as quantum teleportation. This could pave the way to a quantum internet (see Chap. 4).

> Since the beginning of the 2000s, the increasing control of entangled quantum states has led to a second quantum revolution, also known as "Quantum 2.0".[5]

The opportunities offered by this new dimension of quantum physical application inspire the fantasies, not only of researchers, but increasingly those of companies and governments. Since about 2012, many national and international funding programs have been launched. Examples are:

- the Canadian *Institute for Quantum Computing* in Waterloo with initial funding of about $300 million,

---

[5]See also: A. Aspect, John Bell and the second quantum revolution, Foreword to *Speakable an Unspeakable in Quantum Mechanics: Collected Papers in Quantum Philosophy*, Cambridge University Press (2014).

- the Singapore *Centre for Quantum Technologies*, with a starting capital of 158 million Singapore dollars,
- the *Joint Quantum Institute* in the USA,
- the *Engineering and Physical Sciences Research Council* in the United Kingdom, with an investment of 270 million euro,
- *QuTech* in the Netherlands,
- and most recently (in 2016), the EU announced a Europe-wide flagship project for quantum technologies, funded to the tune of one billion euros and due to start in 2018.

Many scientists, manufacturers, and research politicians had been calling for the latter commitment by the governments of EU member states[6]:

Europe needs strategic investment now in order to lead the second quantum revolution. Building upon its scientific excellence, Europe has the opportunity to create a competitive industry for long-term prosperity and security.

Governments do expect a return on their investments. But internet and computer companies are also investing a lot of money in the new quantum technologies. Recall from Chap. 4 the activities of Google, IBM, and Microsoft for the construction of a quantum computer. In addition to the major contenders in this area, many smaller organisations are also getting involved. In many countries including the United States, France, England, and Switzerland, on the basis of their university research, scientists have founded companies that specialize in technologies exploiting entangled quantum systems.

Scientists, politicians, and entrepreneurs have realized that quantum technologies 2.0 constitute a key area of technology for the 21st century.

---

[6]See https://msu.euramet.org/current_calls/fundamental_2017/documents/Quantum_Manifesto.pdf (May 2016), S. 6.

26

Schrodinger's Cat Is Alive: The Path Back to Classical Physics

On their way to developing all these exciting new technologies, quantum engineers had to overcome a major hurdle. For a particular property of the quantum world considerably complicates the control of entangled quantum states: *decoherence*. The reader may remember from Chap. 4 that decoherence refers to the inevitable interaction between a quantum system and its environment, which causes the entanglement and superposition of quantum states to dissolve very quickly.

The significance of the decoherence phenomenon was first recognized in a very different context, not related to the technological exploitation of entangled states, but in attempts to clarify one of the fundamental questions of quantum theory: How does a superposition of different quantum states in the microcosm develop into the well-defined classical states that prevail in the macrocosm? This was precisely the question that had led Schrödinger to his famous thought experiment in 1935. The problem had been around for a long time and was in need of a clear answer. This is what we shall discuss in this chapter. At the same time, it will become clear just how arduous is the task of building a quantum computer in practice.

In the previous chapters, it became clear again and again that quantum objects themselves have no reality of their own, but something that Werner Heisenberg called potentiality. This potentiality only becomes a reality when the quantum object interacts with a macroscopic measuring device. Before the measurement, not only do we as observers not know where and in what state the particle is; even the particle itself does not "know". It can thus move on two paths at the same time and be in two places at once. It is only *with* the measurement that we know (and the particle itself knows) where

© Springer Nature Switzerland AG 2018
L. Jaeger, *The Second Quantum Revolution*, https://doi.org/10.1007/978-3-319-98824-5_26

it is located or in which state it is. Things in the micro world are character-ized by the fact that they are free of definite properties. They do not exist in the traditional sense. In the macro world, on the other hand, we are deal-ing with well-defined and unambiguous characteristics—and that is what we experience in everyday life.

The micro and macro worlds do not seem to fit together properly. For example, how can a table have a real existence and distinct characteristics in the macro world, when it is undoubtedly composed of the smallest, "inde-terminate" particles? Why do not we also experience micro world phenom-ena such as superposition and entanglement in our everyday macro world?

> The key question is: Why do the bizarre properties of the quantum world not appear in our macro world?

This was precisely the question at the centre of the fierce philosophical debate between the fathers of quantum physics. The highlights were:

- the Bohr–Einstein debate (1927 and 1930),
- Schrödinger's thought experiment involving a superposition of a living and dead cat (1935),
- and the EPR paper by Einstein, Podolsky, and Rosen (1935).

Einstein, Bohr, Schrödinger, and their colleagues did not find a satisfactory answer at the time. The "Copenhagen interpretation" suggested by Bohr postulated pragmatically (some would say dogmatically) that a macroscopic measuring system simply obeys classical laws, and microsystems follow their own quantum laws independently of this. The two worlds were separated by "Heisenberg's cut". Most physicists readily accepted this explanation. Only a few objected, including Einstein and Schrödinger.

> With their interpretation that the micro and macro worlds obey completely different laws, Bohr and his followers had prescribed a chill pill to physicists: "That's the way it is, period!"

But this did not sound so convincing. Instead of settling for the "period", about a generation later, some physicists tried to address the concerns expressed by Einstein and Schrödinger and began to look more closely at the transition from the quantum physical microcosm to the classical macrocosm.

The Search for the Elusive Dividing Line

Until then, physicists had considered *either* pure quantum systems *or* unambiguous macro systems, describing the former quantum mechanically and the latter classically. But there is an area in between! And the idea was to explore it by increasing the size of quantum systems in theoretical calculations as well as in experiments. Instead of hydrogen atoms, larger and larger molecules or crystals in solids were gradually considered. Somehow and somewhere the dividing line between the quantum and classical worlds had to manifest itself!

Physicists had dealt with a similar problem decades before: the statistical description of many-particle systems, which some 50 years before quantum physics had become the basis of the physical discipline of thermodynamics. Among other things, *time* had played an important role there. The classical mechanics of a *single* particle is time-reversible, since associated with a given solution of Newton's equation, there always exists a second possible solution that moves backwards in time. The movement of the particle can occur from A to B and forwards in time from time t_1 to t_2, or from B to A and backwards in time from t_2 to t_1. Both paths are solutions of the equations of classical physics. Physicists say that Newton's equation is *invariant under time reversal*.

In contrast, the motion of a large number of particles in an ensemble is no longer time-reversible, neither in theory, nor in practice, as the following example illustrates. A gas of many particles fills a bottle. When the cork is removed. The gas spreads until it fills the newly available space. A single gas molecule can find its way back into the bottle, but never will all the gas particles re-enter the bottle at any given time. The laws of thermodynamics in many-body systems allow only time-irreversible behaviour—the law of entropy (the second law of thermodynamics) applies.

In other words, the motion of a single particle is calculated with very different formulas than the behaviour of a gas of very many particles. In one the motion is time reversible, in the other, it is not. Once again, we can ask: Where is the "tipping point"? At 10 particles? At one million? Or at one million billion?

A diffusing gas provides an analogy with the events in the micro world: there has to be a point at which one set of laws loses its validity and gives way to the other, right?

However, in the case of the gas from the bottle, physicists cannot say at which point exactly the dynamics of the system become irreversible. Such a tipping point cannot be identified. The same is true in quantum physics, as we have seen: there is no unambiguous point, or rather, no clear transition, at which the quantum laws transform into the classical laws.

The search for the dividing line turned out to be a dead end. Only when physicists dealt more closely with the measuring process itself did the breakthrough occur.

When Quantum Systems Mingle with the Crowd

About a generation after Einstein, Bohr, and Schrödinger, physicists realized that they had fallen victim to a fallacy. All too carelessly, they had adopted the concept of closed systems from classical physics and transferred it to quantum physics. What does that mean?

A closed system is a theoretical idealization in which the observed physical system does not interact with the outside world. A single free electron or hydrogen atom, which can be exactly described by the Schrödinger equation, can be considered as such a system. Interactions with other particles from the environment are left out of the equation, if only because their calculation would be far too complicated to be actually carried out with today's means (a quantum computer could help in the future).

But with each measurement, an interaction with the measurement environment *inevitably* occurs. Although most experiments are carried out at temperatures near absolute zero and in the highest possible vacuum in order to eliminate interfering effects as far as possible, it takes at least a single photon to measure the position or the momentum of an electron. This single photon already alters the quantities to be measured. This was what led to Heisenberg's uncertainty principle.

A quantum system subjected to a measurement is no longer a closed system, but an open system. The measurement of a quantum system is therefore always dependent on the measurement environment. Physicists also speak of "quantum contextuality".[1]

[1]Simon Kochen and Ernst Specker, and independently John Bell, were able to prove that quantum mechanics is contextual for systems of dimension 3 and higher. See S. Kochen, E. Specker, The problem of hidden variables in quantum mechanics, *Journal of Mathematics and Mechanics*, 17, 59–87 (1967).

We should therefore no longer wonder about the measurement problem, but rather about physicists' endless attempts to transfer statements about isolated, intrinsically indeterminate quantum systems to our everyday lives. For a pure, and in principle reversible quantum process could only take place in a system that has detached itself from the rest of the universe and left no trace in it.

And now comes the crucial point. Of course, it is not only measurements deliberately planned by humans that prevent quantum particles from behaving completely undisturbed and thus displaying their typical quantum properties, such as superpositions of states. Many of the contradictory phenomena of the quantum world, such as the double-slit experiment and particle entanglement, are extremely sensitive to *any* environmental impact: collisions with gas molecules, but the emission or absorption of radiation ("collisions" with photons) also quickly destroy the relationships within the quantum mechanical wave functions (their so-called phases) responsible for quantum phenomena. Even in outer space, where individual particles or atoms are positioned relatively far apart from each other, they are permanently disturbed, for example, by the cosmic background radiation. Physicists also speak of the "thermal bath" of the environment.

The environment, in a sense, permanently carries out measurements, and even without humans as observers, individual quantum systems can only ever be approximately regarded as "pure quantum states".[2]

Reality by Decoherence

Upon realizing that they almost always had to deal with open systems when investigating quantum systems in practice, physicists finally found a way to understand and eventually describe the dynamics of the transition from quantum states with their superpositions and entanglements to unambiguously defined classical states. In the early 1970s, the German physicist Dieter Zeh

---

[2]For those readers who would like to go a little deeper, the above analogy with a many-body system in statistical physics (thermodynamics) can be given a deeper meaning here. The interaction between quantum particles and their (macroscopic) environment also means that the second law of thermodynamics (entropy theorem) comes into play. While the dynamics of a quantum particle is reversible (like Newton's equation, the Schrödinger equation is also reversible) the interaction with an open environment is not. We therefore have to consider the irreversible effects of the measurement environment, and these eventually paved the way to a coherent interpretation of the measurement process in quantum systems.

achieved an important breakthrough with the development of the concept of decoherence. Then in the 1980s, physicists were able to develop a consistent mathematical description of this concept.[3]

Generally, decoherence is an essential feature of the dynamics of open quantum systems. We have already encountered this concept as one of the major problems in the design of quantum computers (see Chap. 4). Here quantum-mechanical superpositions of as many states as possible have to be maintained for a sufficiently long period of time to perform the desired calculations. However, so far, due to external disturbances, these superpositions decay too quickly.

The embedding of a quantum system in a larger whole makes it necessary to explicitly include the environment in the mathematical models. The theory of the measuring process then looks like this:

- It starts with the contact between the microscopic quantum system with its superposition states and the macroscopic measurement environment.
- The interaction that occurs as a consequence of that contact leads to a quantum mechanical entanglement between the originally isolated states of the quantum system and the (many) degrees of freedom of the classical measuring system, leading to a single large total wave function (hereafter called the "total system").
- The original quantum system is therefore no longer in single and pure states, independent of the measuring system; its states are part of a "statistical mixture" of all states in the total system. The states of the two systems are thus all linked together in a complex way and can no longer be viewed separately. In mathematical jargon, the quantum system and the measuring system no longer "factorize" into a product of separate states.
- The information about the states of the original microscopic system is now in the wave function of the total system, which represents an inseparable mixture of all states, i.e., the measurement system together with the originally isolated quantum system.
- The relationships between all the participating states, including superpositions and entanglements between the quantum system and the states of the measuring system, are initially preserved. One could say that, for a very short lapse of time, there actually exists a macroscopic superposition of dead and living cat, just as there is a superposition of decayed and non-decayed atomic nucleus.

---

[3]For a detailed reference see D. Zeh, *Decoherence and the Appearance of a Classical World in Quantum Theory*, Heidelberg (2003).

Upon measurement, the micro and the macro worlds become entangled, creating new superpositions. But now they are superpositions of all components with each other, i.e., those of the quantum system and those of the measuring system.

So far so good. This is all well known within the formalism of quantum mechanics. However, if we now consider (at least conceptually) once again the state of the original quantum system, but *within the total system*, i.e., together with the connected (entangled) states of the measuring system, something amazing happens. The temporal development of the total system (in the language of physicists, "unitary" development[4]), including the interaction between the quantum system and the many degrees of freedom of the measuring system, leads to the bizarre quantum properties of the individual particles, which fit so poorly into our everyday perception, averaging out and disappearing. Physicists say that the interference terms in the wave function disappear. In the calculation for the total system, this leads to an irreversible resolution of all superpositions in the original quantum system.

The "coherent" properties of the original quantum system, i.e., the (phase) relationships of its wave function, which are pivotal for the occurrence of superpositions and interference effects, and their correspondences in the measurement system are thus irretrievably destroyed within the wave function of the overall system during the measurement process. Likewise, any entanglement with the wave function of the measuring system decays almost immediately, and with this the superpositions of the states of the various particles within the total system. This ultimately leads to the classically measurable states of the quantum system. The whole process is irreversible in the sense of the entropy theorem.[5]

One can also view the irreversibility of the measuring process and the destruction of the superpositions from the perspective of entropy and the information generated by the measurement. According to the laws of thermodynamics, the generation of information associated with the measurement process always comes with a corresponding increase in the entropy of

---

[4]"Unitary" is not a translation of "temporal", but a mathematical property of the operators for temporal evolution.

[5]The irreversibility is here defined (as it is in thermodynamics) in such a way that small errors in the reversal of the particle motions (in the classical gas, the impulses of the particles) lead to very different final states, when time is then reversed. So this is an example of (quantum) "chaos". With perfect inversion, the previous system would reappear.

the system (in fact, information and entropy are mathematically equivalent to each other). With the inevitable increase in entropy, the measurement becomes irreversible, and as a consequence the wave function decays.

> The paradoxical quantum realities disappear almost instantaneously at the macro level; with the measurement, each microsystem is quickly transformed into its ordinary, classical counterpart with single, unambiguous states.

Indeed, the process of decoherence occurs upon *every* interaction with a macroscopic system. So we do not have to wait until a concrete physical measurement, or even a conscious being such as "Wigner's friend", interrupts the in principle infinite sequence of measurement processes. This already happens as soon as the quantum system interacts with any macroscopic system—which it inevitably and permanently does when embedded in a classical environment. One could also say that the complex interactions between a quantum object and its macroscopic environment conceal the quantum effects from us.

> Before we open the door to the box containing Schrödinger's cat, the environment has already made, so to speak, billions of observations that have long since destroyed every possible superposition of decayed and non-decayed particles, or indeed, living and dead cats.

Ultrafast Transition to Reality

The "collapse of the wave function" during the measurement process which Einstein deemed so ominous does actually take place, albeit not completely spontaneously, but in an ultra-short lapse of time called the "decoherence time". The disintegration of macroscopic superpositions within the total system is in fact a process that can be entirely described within quantum theory, and not, as in the Copenhagen interpretation, a process that requires an outside framework, as it were. The decoherence time for free electrons in an ultrahigh vacuum is about 10 s, long enough that its particular quantum properties can be measured. For a dust particle of 10 micrometre it is only 10^{-4} s, and with a football the superposition states have already disintegrated after 10^{-12} s. (This is the value for an ultra-high vacuum. At normal pressure, the decoherence times are much lower still: for footballs at room temperature, they are about 10^{-26} s).

In macroscopic systems the classical notion of uniqueness of states (and as a consequence, the separation of subject and object) are valid for all practical purposes, while in the microcosm this approximation collapses.

We do not therefore need two separate worlds with two different theories and concepts of reality. As suggested in the Copenhagen interpretation, physicists may treat the measuring instrument and the quantum system separately, the former classically, the latter quantum physically. However, while Bohr talked somewhat vaguely about a "complementarity" between classical and quantum mechanical laws, and thus presumed classical physics as necessary a priori for the measurement process, the principle of decoherence offers the possibility of describing the measurement process entirely *within* quantum theory. It brings the two worlds together without the need for further assumptions.

In the last few decades physicists have therefore overcome the idea of Heisenberg's cut: quantum theory, which includes interactions in an open system, is a theory of the microcosm *and* the macrocosm, and thus a conclusive and complete physical theory. As the decoherence time of a football shows, quantum theory is universally valid, but it is largely irrelevant for the macro world. Superpositions of quantum objects—for example, the combination of decayed and non-decayed particles (or cats)—are in principle also possible in macroscopic systems, but in practice, in systems with many particles, there is so much interference that superpositions of macroscopic states are barely ever observable.

The macro world including the measuring process can be completely described within quantum theory. Its laws apply in both worlds, the microcosm as well as the macrocosm.

The Next Steps

However, the question of the nature of the measurement process has still not been given an entirely satisfactory answer yet. If in quantum mechanics the individual components of a system are connected (entangled), if their states can represent superpositions of several classically possible states, even if one can no longer speak of independent components of a system, it remains open until today how *exactly* the measurement process dissolves the superpositions and how *exactly* this process can be described. This is because the decoherence concept does not explain individual measurements but only makes statistical statements about many measurements.

- Why, for example, is exactly one of the many possible classical states realized? Which selection mechanism is responsible for picking out a particular state in a fluctuating quantum system?[6]
- If decoherence in the measurement makes the quantum properties of the system disappear, where *exactly* is the transition between the measurement device obeying classical physics and the object of measurement which follows quantum physical principles? This question stands in analogy to thermodynamics and the statistical description of multi-particle systems.
- Where exactly is the boundary between the bizarre entanglement of all components in the microcosm and the separability of objects in classical physics?

These questions are keeping physicists busy today. Recently, they have succeeded in observing single quantum particles without destroying them, i.e., they been able to witness their superpositions, observe the decoherence process directly, and investigate it experimentally. For such experiments, Serge Haroche and David Wineland were awarded the Nobel Prize in Physics in 2012.

And there is another question that provides food for thought for physicists: despite decoherence, should quantum phenomena not in principle also be observed in macroscopic systems? Decoherence explains why quantum phenomena can not *normally* be observed in the macro world. Nevertheless, as has been clear from the last few chapters, quantum effects can also be found in macroscopic systems. The laser, Bose–Einstein condensation, and superconductivity and superfluidity at very low temperatures all rely on macroscopic quantum effects, where a large number of particles can be described collectively by a *single* wave function. The quantum-mechanical effects of entanglement and superpositions thus occurring cannot be described by the laws of classical physics. The size of a physical system alone does not exclude quantum effects per se.

---

[6]The theoretical physicist and expert on decoherence W. Zurek even speculates that principles of Darwinian evolution may apply here.

By connecting the problem of measurement with the phenomenon of decoherence, we are just beginning, almost 100 years after the birth of quantum physics, to reconcile it with the macroscopic world of our experience.

The Flip Side of Decoherence

Long ignored by the general public, quantum physics has made tremendous progress in recent years. Physicists have learned to forego descriptive explanations for atomic systems and to accept the bizarre consequences of quantum theory. Decoherence has given them new answers to some fundamental physical and philosophical questions concerning the quantum world and has finally explained why we do not equally experience these consequences in the macrocosm. At the same time, decoherence poses great challenges for physicists. On the one hand, the dissolution of entanglement and the decay of superpositions produce the macroscopic states of our world. On the other hand, the same processes make entangled states so short-lived that their use in new quantum technologies remains highly challenging. Especially for the construction of quantum computers, the fast decoherence of quantum states provides one of the main challenges.

By describing the decoherence of quantum states, physicists have found the pivotal step towards unifying the microcosm with the macroscopic world. At the same time, decoherence represents the key hurdle for new quantum technologies, such as the quantum computer.

This new knowledge should find its way into the philosophical debate about the interpretation of quantum physics. The discovery of decoherence means that we no longer need to resort to any external entities, such as human consciousness, to explain what is happening in the micro world. Bohr and Einstein would probably have enjoyed seeing how, thanks to decoherence, the peculiar properties of the quantum world can be consistently reconciled with the world on our own scale of length and time. And Schrödinger would certainly have been happy to learn that he would not be responsible for the death of a cat solely by opening a door.

Part VI

The Future—Where Are We Going?

27

Quantum Revolution 2.0: When Nanobots and Quantum Computers Become Part of Our Everyday Lives

The previous 26 chapters have shown that quantum theory represents the greatest scientific makeover of the 20th century. Furthermore, the fact that we live in a world that is *only seemingly* real and deterministic represents a complete break with our everyday habits of thought. We still do not know how this realization will affect our future thinking. The philosophical meaning of a dissolution of the subject–object dualism in the microcosm, the principles of symmetry in theoretical physics, and the non-local effects of entangled particles have not yet penetrated deeply into our everyday lives and thinking. And yet quantum physics with its insights has already fundamentally shaped our modern worldview. Today many people have said goodbye to the comfort zone of absolute certainties, be they of religious, philosophical, or even scientific nature. They can live well with the ambivalence of complementary truths (in the sense of Bohr[1]). This is not the least merit of quantum theory.

What else can we expect? In the past, great upheavals in our worldview have have sooner or later always fundamentally changed our lives:

- The first example, historically speaking, is the emergence of rational philosophical thought in ancient Greece. People were no longer satisfied to provide traditional (religious) answers to the fundamental questions of humanity, like how the world came into being, what happens to us

---

[1]This is in no way comparable to the recently popularized "alternative facts". These are nothing but lies.

© Springer Nature Switzerland AG 2018
L. Jaeger, *The Second Quantum Revolution*, https://doi.org/10.1007/978-3-319-98824-5_27

after death, why this or that natural phenomenon occurs, and so on. The image of the supreme god Zeus sending flashes of fire down to Earth was no longer good enough; world affairs became increasingly subject to critical questioning, based on the laws of logic and the criteria of empirical observation. This "transition from myth to logos" took several centuries (from about 800 to 200 BC). The mixture of a naturalistic and a rational perception of nature which was born during this time continues to shape the way people think today.[2]

- Then there came the development of the scientific method in the late Renaissance. After one and a half millennia of religious orthodoxy, people rediscovered the thinkers of Ancient Greek and began once again to assess nature empirically and rationally. What was new was that scientists now tried to describe nature systematically and theoretically with the help of mathematical laws. This led to radical changes in our philosophical, theological, social, and political understanding. Humans soon felt that they were no longer at the mercy of nature. Their desire for an individual way of life, economic autonomy, and the discovery of new horizons superseded the intellectual and geographic narrowness of the Medieval Age. And from the scientists' attempts to understand the world came a growing desire to reshape it.

- A new, critical form of scientific thinking became prevalent during the Enlightenment era. In Newton's mechanics, God was left with the role of the watchmaker. Suddenly, there was no longer an eternal "God-ordained" order and the religiously substantiated legitimacy of political, social, and economic power began to erode. Over thousands of years, impenetrable barriers between hierarchical social structures gradually became permeable. All this contributed to a much greater human intellectual potential—we would say "human capital" today. Born in the early 17th century, Albert Einstein would most likely have become a simple merchant like his father. In the 20th century he was able to revolutionize our worldview as a physicist.

- Darwin's theory of evolution moved man away from the centre of creation, making him the result of a process that all animals and plants had also been through. As a result, God as the Creator and other such

---

[2]Interestingly, the monotheistic and transcendent religions also came into existence during this period. For more on this relationship see (in German) L. Jaeger, *Wissenschaft und Spiritualität*, Springer Spektrum, Heidelberg (2017).

transcendent principles became permanently superfluous.[3] Darwin's statement that each human being is evolutionarily unique did much to fuel the pronounced individualism of the modern world. The new image of man also had an impact on moral values: the widespread thesis of social Darwinism placed self-preservation and one's own personal success at the centre of human aspiration. Darwin's principles soon escaped from their exclusive reference to physical survival and biological reproduction and were also applied to the social and political fabric of human existence.

We may predict that the discovery through quantum theory that our world in its microstructure is non-real and nondeterministic will further revolutionize millennia-old principles of our lives and the way we understand ourselves. The changes we have made so far in our self-perception are probably only the harbingers of far more radical changes.

> The discovery of quantum physics was the most important philosophical event of the twentieth century and is likely to change our worldview even more drastically than it has done so far.

Technology and Social Change

Just as certainly as the increase in scientific knowledge entails major ideological upheavals, so it has always had a major impact on technological, social, and economic developments. In fact, the natural sciences are the main driving force for our modern prosperity. The relentless human pursuit of knowledge gives rise to *scientific* progress which, combined with the dynamism of free-market competition, results in equally steady *technological* progress. The first offers humankind ever deeper insights into the structure and processes of nature, while the second provides us with almost limitless possibilities for private pursuits, economic development, and improvements in the quality of life.

---

[3]While the scientific revolution of the sixteenth and seventeenth centuries was not yet a fundamental rebellion against fidelity to the Christian faith, the new treatment of the fundamental questions of biology and the characteristics and evolution of life on Earth necessarily depends on whether one refers to an act of God when asking about the beginning of life. Biology and Darwin's theory of evolution therefore penetrated far more deeply into the principles of religious conviction than did the discoveries of physics during the scientific revolution of the 1700s. See also (in German): L. Jaeger, *Die Naturwissenschaften: Eine Biographie*, Heidelberg (2015).

Here are some classic examples:

- New technological discoveries made during the Renaissance, like paper-making, letterpress, mechanical clocks, navigation tools/shipping, construction, and so on, all brought unprecedented prosperity to Europeans.
- The Industrial Revolution of the 18th and 19th centuries saw the fruits of Newtonian physics and found a dramatic technological manifestation in the form of steam engines and heat machines, based on the new theory of heat. Railway and industrial machinery revolutionized transportation and production. For a large part of the rural population, the relocation to cities and the new work environment in factories brought a completely radically new life compared to their ancestors.
- Faraday and Maxwell's electromagnetic field theory in the late 19th century led directly to the electrification of cities, modern telecommunications, and electrical machines. The new possibilities of rapid intercontinental communication and the new transport system resulted in a first wave of political and economic globalization.
- The technological revolution of the twentieth century basically corresponds to the first generation of quantum technologies and has brought us lasers, computers, imaging devices, and much more (including, unfortunately, the atomic bomb), which have all become part of our modern day lives. Digitization, with its ever-faster processing and transmission of information, the integration of production with information and communication technologies, and of course the internet, has produced another powerful wave of political and economic globalization.

The upcoming second quantum revolution will create something new again. It will once again completely revolutionize communication, interaction, and production.

> Like all previous technological revolutions, the Quantum Revolution 2.0 will result in yet another dramatic change in our way of life and society.

The Mighty Trio

All in all, there are three focal technological areas which will each shape our society profoundly in the near future:

- genetic engineering,
- artificial intelligence (AI), and
- quantum technology 2.0.

Gene technology and artificial intelligence are widely regarded as threatening, and the debate over their use and impact is in full swing. In fact, both technologies have the power, not only to change our everyday lives, but even humankind itself.[4] For example, they could in the future be used to merge humans and machines in order to potentiate our capabilities by combining our cognitive abilities with the computational and physical performance of machines. However, machine intelligence superior to our own even in general thinking abilities is also conceivable, and this not just in arithmetic, chess, or Go.

However, quantum technologies 2.0 (including quantum computers and nanomaterials) so far appear merely as fuzzy points on the radar screens of those concerned with the impact of new technologies on society. At the same time, the three technologies mentioned above can hardly be separated from one another. They will cross-fertilize each other and together generate a much higher impact. For example, the potential of new quantum technologies could give AI and genetic engineering a significant boost:

- The computing power of quantum computers could once again massively improve the optimization algorithms for neural networks in AI research.
- Nanomachines could replicate themselves according to a manual given by humans and improve these instructions on their own through genetic algorithms.
- Smart nanobots working as a genetic editing engine could actively manipulate our DNA to permanently repair and optimize it. The only problem is to decide whose ideas will be used to determine what comprises an optimization.

The impact of Quantum Technology 2.0 is largely underestimated. Just its contribution to the development of artificial intelligence and its potential use in genetic engineering will be of great importance.

---

[4]For a more detailed discussion see (in German): L. Jaeger, *Supermacht Wissenschaft*, Gütersloher Verlagshaus, Gütersloh (2017).

The focus on new quantum technologies is today still restricted to a discussion about the potential health hazards of nanoparticles in our bodies. This strange dismissal of the power of quantum technology is not entirely harmless. For this blind spot is compounded by another perceptual bias: we have become accustomed to the fact that technological progress is getting faster and faster, but we underestimate its *absolute* speed. An example of this is Aldous Huxley's famous 1932 novel *Brave New World*.

Fast New World

In his seminal book Huxley describes an eerie futuristic scenario about a human society made up of numerous castes of genetically manipulated humans. Due to the genetic manipulation, everybody's social status is already determined at birth; the hierarchy includes five classes of humans, from alpha to epsilon. Alpha humans form the leadership caste, while the intelligence of the Epsilons, only used for simple tasks, is artificially reduced to a minimum. Huxley sets his scary scenario take place in the year 2540, because he reckoned it would take more than 600 years for such a scenario to become technologically feasible (the *social* acceptance of such a world did not appear as far-fetched in the 1930s). However, modern gene editing methods make this scenario look much more feasible today from the technological point of view, less than 90 years after the book was published.

> Aldous Huxley set his dystopia *Brave New World* 600 years in the future. However, an implementation of the scenario he described appears technologically possible less than a hundred years after its publication.

Many future scenarios from the last hundred years or so are no longer science fiction fantasies today. For all the technologies mentioned in the following, the scientific basis is currently being developed in laboratories around the world. Here is a selection of quantum technological developments from the last 26 chapters:

- Health: Nanobots will be used as super-small tracking devices and molecular robots. They will move around inside the body, where they will recognize and treat cancer cells, vessel plaques, and pathogens at an early stage.

- Enhancement of mind and body: Artificial body parts made of nanoparticles, for example an artificial nano retina, will be able to improve our sensory perceptions and physical abilities. Brain chips will increase our cognitive and communication skills.
- New dimensions of artificial intelligence: "Quantum Machine Learning" will combine quantum mechanics with the latest machine-learning techniques to develop an artificial intelligence that will exceed human cognitive abilities in a way that we human beings will no longer be able to comprehend.
- Production of goods: A "quantum 3D printer" will be able to arrange individual atoms in almost any imaginable way—for example from a handful of dust—at the touch of a button or even by thought control. Through this targeted atomic arrangement, matter can be brought into completely new forms and functions. Programmable, intelligent materials will characterize our everyday life, just as plastic cups and metal devices do today. A future advertising slogan could be:

You do not like your apartment any more? We can program a new one for you within a day.

- Economics: If matter can be manipulated almost without restriction—for example, by printing food or programming it to take on almost any properties—everyone will immediately get what they want. The lack or shortage of goods and resources would have a huge impact on the economy and society as a whole. What would an economy look like in which ownership no longer matters? What jobs would have to be done? Would everyone then really be socially equal?

Future quantum technologies will fundamentally change our ideas of personal possessions and social status, of health, and finally of ourselves, for example through a kind of coalescence of man and machine.

All these exciting, auspicious, and at the same time frightening possibilities of future quantum technologies (as well as all other technologies) raise many questions:

- Will we be able to control the unlimited computing power of quantum computers?

- What happens when an artificial intelligence develops that is superior to us, not only in *some* cognitive domains, but in *all* domains?
- And do we really want our brains to connect to nanobots?

The fundamental challenge will be to find the answers to the following questions: How can the expected technological progress be designed so that it does not overrun us? And how can we cater for the expected social tensions?

> If it is a frightening thought that we could control the future of humankind itself and our society through Quantum Technology 2.0, Genetic Engineering, and AI, then the scenario that we have this technological power and *cannot* control it is much worse.

How we deal today with the issues that arise with technological progress on an ethical and social level will determine the future of our individual dignity and freedom, and ultimately of humanity as such. But who could undertake the task of directing our knowledge and technological creativity along socially acceptable paths?

Who Runs the Show?

Several social actors come to mind which could potentially guide technological progress in a way that would be compatible with our human values. However, two of the most often mentioned social players would undoubtedly be overwhelmed if they were the sole such designers:

- The responsiveness of social decision-makers (politicians, business leaders, media workers, etc.) whose job it is to increase the common good is far too slow to handle the ever accelerating dynamics of technological change. Among other things, this is because our political, economic, and cultural leaders have little in-depth knowledge of the current state of scientific research.
- Scientists themselves will be just as unable to control technological progress. Quite the opposite, in fact. Like all other members of society, they largely operate under the logic of the market. They could become billionaires today, if they develop new technologies based on their insights. Moreover, they are always dependent on the government or other institutions to allocate their research funds.

A third socially creative force is the free market. So far technological progress has almost exclusively followed the logic of a market-based (or military) utilization. In other words, whatever was possible and gave a financial (or military) advantage to somebody was implemented. Can we hope that the mechanisms of free market competition will control technological progress in the best possible way for the common good?

> Leaving progress to the free market would mean that Google, Facebook, and Amazon would determine the use of quantum computers or higher artificial intelligence.

That this would turn out well for all of us might well appear far-fetched, even for the most devout supporters of the free market ideology. In fact, the market is a very poor arbiter when ethical issues are involved. In order to decide how far we can leave the development of future technologies to the free market, we need to know and name the forces that keep it from making the best decisions for society as a whole. Apart from the prospect of billions of dollars of business, which alone would lead to almost insurmountable conflicts of interest, there are other problems with putting unconditional trust in the forces of the free market:

1. Externalities: The economic activities of one group may have an impact on other groups—possibly even on all people worldwide—without the actors paying the full cost for it. Such externalities are particularly evident in public goods that have no market price. These include, among other things, ecological resources and overall health. Some examples are:

   - polluting the environment still costs little or nothing to the polluter,
   - climate-damaging CO_2 emissions are still not associated with higher costs for producers,
   - the safety risks associated with nuclear power generation or natural gas fracking are largely borne by the general public, and
   - while the massive use of antibiotics in agriculture produces higher yields for agribusinesses, it is enabling resistant bacteria to become a global health risk.

2. Rent-seeking: Powerful groups often succeed in changing the political and economic rules to their own advantage, thus receiving various forms of state guarantees without increasing the overall social well-being, and often even diminishing it. The most obvious example is corruption.

3. Information asymmetries: In 1970, in his essay "The Market for Lemons", economist Georg Akerlof showed that free markets cannot function optimally unless buyers and sellers have the same access to information. However, in many markets, we can observe significant asymmetries in information access: the labour market, the market for financial products (this is what lets banks get away with excessive fees for their investment products), the healthcare and food markets, the energy market, and, most importantly in our context, the market for new scientific knowledge and technologies. Anyone who wants to weigh up a new technology against its risks needs to understand it in detail. But the one who knows most about it is its inventor and manufacturer, who will generally be less interested in its risks than in opportunities for profit. For profit-oriented companies in a free market economy, lying is just part of the game. This includes the systematic sowing of doubt about established scientific knowledge. Incidentally, in 2001, Akerloff was awarded the Nobel Prize for Economics for this insight.

4. Cognitive Distortions: Standard economic theory assumes that we know what is good for us. Yet behavioural economics has long since shown that we often act far less rationally than the advocates of the free market would have us believe. Thus, producers and consumers are often guided by short-term emotional drives rather than by long-term rational considerations.

These are four reasons why the free market is unsuitable for guiding technological progress in a socially acceptable way.

> The capitalist logic of exploitation is a tremendous force that counteracts differentiation and ethical reflection in the development and judicious use of new technologies.

Informing Ourselves Is Our First Civic Duty

One thing is certain: the world will change fundamentally as a result of future quantum technologies. That is why our decisions today have tremendous leverage. Just as the scientific foundations of contemporary automobile, train, and air traffic were laid in the eighteenth and nineteenth centuries, and that of modern communication and data processing in the nineteenth and twentieth centuries, the foundations for the wonder technologies of the twenty-first century are being built today. There remains only a very small time window before technologies and social norms will have become so well established that we won't be able to go back. This is why an active, broad social and, of course, democratic dialogue is so urgently needed.

> The ethical evaluation and political shaping of future technologies must go beyond the commercial or military interests of individuals, companies, or states.

This will require a democratic commitment from each of us, including the duty to inform ourselves and exchange views and perspectives. It should also be our demand on the media that they provide comprehensive information on developments and progress in science. There is far too little talk about physics, chemistry, or biology when journalists and others who guide public opinion report on world affairs and important social developments.

Plus, in addition to ethical integrity we must demand from politicians and other societal and economic decision-makers an attitude of *intellectual* honesty. This means that deliberate falsehoods, but also distortion of information and filtering of information for the purpose of enforcing particular interests, must be consistently combated. It is unacceptable that fake news should be able to unfold its destructive propagandistic power in the way it does these days, and that such a frightening number of politicians, for example, should still seriously doubt climate change or Darwin's theory of evolution.

But the commandment of intellectual honesty applies equally to the recipients of information. We have to learn to take time before we draw conclusions, to analyze our own prejudices, and to engage in complex interrelations without always oversimplifying. And lastly, we have to allow for inconvenient truths.

> The task of every citizen is to strive for a broad, rational, information- and fact-based discourse in shaping our technological future.

It will pay to follow the evolution of quantum physics research very closely. We are experiencing a defining moment in human history, where the peculiar properties of the quantum world are becoming an integral part of our everyday lives. Those who do not look closely run the risk of missing out and only realizing what has happened when it's too late. Our present understanding of the phenomenon of entanglement gives us a glimpse of what may be possible in a seemingly distant technological future. But it is a future that has already begun.

Epilogue: One Morning in 2050

Markus, born in 2020, sleeps a bit longer today. It is his 30th birthday. His fMRT alarm clock connects directly to his brain, logging into his dream, letting it become lucid (in lucid dreams, the dreamer is aware that he is dreaming) and thus communicates to Markus's subconscious that it is time to get up. The system has long since calculated the optimal wake-up time, so that Markus awakes from the REM phase as fresh as possible. Shortly before he wakes, the nanobots in his body measure the latest developments on possible inflammations, vessel plaques, or cell mutations. As soon as Markus opens his eyes, the data is displayed on his nano-retina.

As every morning, his breakfast consists of a butter croissant and jam. Again, nanobots have been active. All superfluous sugar and fat molecules have been removed and replaced with valuable vitamins, trace elements, and dietary fibre. The fact that the croissants still taste as buttery as in his grandmother's time, forty years ago, can also be put down to the skills of the nanobots. They stimulate Markus's taste buds with the appropriate neuro-signals. The kitchen is quite empty. Things like kitchen appliances and supplies are no longer necessary. What was yesterday a small oven, perfectly adapted to the size of the roast, has today turned into a toaster. This is made possible by using programmable matter consisting of nanoparticles. Markus slowly spreads the almost fat-free butter on his croissant.

Through his retina implant and a microchip in his brain, Markus is directly connected to the internet, which sends messages tailored to his interests directly into his brain. The AI operating on quantum computers, trained and customized for him and his personality, knows more about Markus's preferences than he does himself, ranging from his favourite football club to his political views. For it has recorded every data point of his life and continually runs algorithms for optimizing his well-being. The communication between Markus and his AI naturally goes both ways. Through his thoughts, he indicates that he wants to know more about the war in the Middle East. He immediately receives the desired information, sent by appropriate signals to the appropriate neurons in his brain, so that he not only sees the war, but also smells, tastes, and hears the smoke and gunfire.

On the wall he recognizes the rainforest landscape, under whose impression he fell asleep yesterday. He still has the smell of humidity in his nose, or rather, in the appropriate neurons in the olfactory bulb of his brain. This morning he prefers a beach, so he makes his wish. Immediately, a tropical coral reef opens up in front of him, filled with the sounds and the smells of the ocean. Or rather inside him, for the perceptions are generated directly in his brain.

When he connects with his girlfriend Iris via *Brainchat*, the new brain-to-brain software, his AI shows him that an unauthorized person is eavesdropping on his quantum communication channel. The software offers to change the encryption or to use a different channel.

The news item played into his consciousness has changed. Now he is following a discussion about the abolition of money. In recent years, the importance of ownership has changed completely. There are no more scarce goods that are worth having money for. Everything material can be created with 3D printers from the simplest basic materials. By suitable neuro-stimulation, all desired feelings and sensations can be produced directly in the brain. The representatives of the new socialist movement demand free access to all software for printing and converting products. The last remaining software companies from the information age back in the first 20 years of the 21st century, *Alphabet* and *Dodax* (the latter had been created in 2029, after the merger between Facebook and Microsoft), continue to resist. But their cause has long since been lost. The free market economy is no longer relevant. Everything people need is available in the form of software. All they have to do is print stuff out or import the appropriate software into the material objects.

In the past, software had required special machines called computers. They were expensive and inflexible. But then, ten years before, when the problem of decoherence of entangled quantum systems had been solved technically, software for quantum computers was developed and implemented into things directly, for almost any form of matter. Quantum computers made it possible to control the individual atoms in a material compound in such a way that they could be put together to form any energetically possible structure. The only thing needed was the appropriate software.

The *New Socialists*, who emerged from the Social Democratic movement in 2041, now have a two-thirds majority in parliament. According to their election platform, they will make free access to all software a constitutional right for every citizen. It would be the end of Alphabet and Dodax. But that would not be such a bad thing, so constitutes the tenor of the present debate. It would be like the extinction of the last dinosaurs.

Markus turns to his hobby, the genetic construction of new animal and plant species. He has not held a paid job for several years and most of his friends are no longer employed either. Just about everything (and probably soon literally everything) he needs is available at the touch of a button. AI-equipped machines and nanobots take care of almost everything. There is simply no reason to earn money anymore. The principle of money as a medium of exchange has lost its meaning, and the next generation will hardly understand why money was once so important. With a shudder Markus thinks back to earlier times when he had to think hard about whether he could buy the latest model of electric car and had trouble repaying his loans. As he bends over his little CRISPR device, he briefly wonders whether his brain chip, which connects him to the central AI, may have been programmed for him to have such an aversion to earlier times. But then he smiles to himself and goes back to thinking about the orange tone of the moss he wants to cover his walls with.

Literature

Aaronson S (2013) Quantum computing since democritus. Cambridge University Press, Cambridge

Akerlof G, Shiller R (2016) Phishing for fools. Princeton University Press, Princeton

Al-Khalili J, McFadden J (2015) Life on the edge: the coming of age of quantum biology. Black Swan, Richmond

Aristoteles (1924) In: Ross WD (ed) Metaphysics. Clarendon Press, Oxford (reprinted 1953 with corrections)

Barnes J (1982) The presocratic philosophers. Routledge, London and New York

Bell J (1987) Speakable and unspeakable in quantum mechanics. Cambridge University Press, Cambridge

Bins C (2010) Introduction to nanoscience and nanotechnology. Wiley, Hoboken

Bohr N (1931) Atomtheorie und Naturbeschreibung (Atomic theory and the description of nature). Springer, Berlin

Bohr N (1958) Atomic physics and human knowledge. Wiley, New York

Boltzmann L (1908) Vorlesung über Maxwells Theorie der Elektrizität und des Lichtes (Lectures on Maxwell's theory of electricity and light). Barth, Leipzig

Born M (1957) Physik im Wandel meiner Zeit (Physics through time). Vieweg, Braunschweig

Capra F (1975) The tao of physics. Shambhala Publications, Boulder

Capra F, Luisi P (2014) The systems view of life: a unifying vision. Cambridge University Press, Cambridge

Carroll S (2016) The big picture. Dutton, Boston

Clayton P, Davies P (eds) (2006) The re-emergence of emergence. Oxford University Press, Oxford

© Springer Nature Switzerland AG 2018
L. Jaeger, *The Second Quantum Revolution*, https://doi.org/10.1007/978-3-319-98824-5

Dalai Lama (Tenzin Gyatso, the Fourteenth Dalai Lama (2005)) Essence of the heart sutra, Jingpa GT (ed). Wisdom Publications, Boston

Dalton J (1808) A new system of chemical philosophy. B. Bickerstaff, Strand and London

Descartes R (1670) Meditationes de prima philosophia (Meditations on first philosophy, translated by Cottingham J). Cambridge University Press, Cambridge (1996)

de Padova T (2013) Leibniz, Newton und die Erfindung der Zeit (Leibniz, Newton, and the invention of time). Piper, München

Doudna J (ed) (2016) CRISPR-CAS. Scion Publishing Ltd, Banbury

Doudna J, Sternberg S (2017) A crack in creation: gene editing and the unthinkable power to control evolution. Houghton Mifflin Harcourt, Boston

Dukas H, Hoffman B (1954) Albert Einstein: the human side—new glimpses from his archives. Princeton University Press, Princeton

Eccles J (1994) How the self controls its brain. Springer, Berlin

Einstein A (1979) Aus meinen späten Jahren (My later years). DVA, Stuttgart

Einstein A (2008) The world as I see it. Citadel Press, New York (original edition in German 1931)

Einstein A (2010) The ultimate quotable Einstein. Princeton University Press, Princeton

Eschbach A (2014) Lord of all things. AmazonCrossing, Seattle

Farmelo G (2009) The strangest man: the hidden life of Paul Dirac, quantum genius: the life of Paul Dirac. Faber and Faber, London

Feynman R (1990) QED, the strange theory of light and matter. Penguin, London

Feynman R (1992) The character of physical law. Penguin, London

Feynman R (1997) Surely you're joking, Mr. Feynman! (Adventures of a curious character). W.W. Norton & Company, New York

Feynman R (1999) The pleasure of finding things out—the best short works by RP Feynman. Perseus Books, New York

Ford K, Wheeler JA (2000) Geons, black holes, and quantum foam: a life in physics. W.W. Norton & Company, New York (revised edition)

Friebe C et al (2015) Philosophie der Quantenphysik (Philosophy of quantum physics). Springer, Heidelberg

Fritzsch H (2005) Elementary particles: building blocks of matter. World Scientific, London

Galilei G (1638/1960) Il Saggiatore (Rome, 1623) (English translation: The assayer. In: Drake S, O'Malley, CD (eds) The controversy on the comets of 1618). University of Pennsylvania Press

Gilder L (2009) The age of entanglement. Vintage, Toronto

Gisin N (2012) Quantum chance—nonlocality, teleportation and other quantum marvels. Springer, New York

Good IJ (ed) (1961) The scientist speculates. Heinemann, London

Görnitz Th (2016) Von der Quantenphysik zum Bewusstsein – Kosmos, Geist und Materie, Springer, Heidelberg (English version: Quantum theory as universal theory of structures—essentially from cosmos to consciousness; to be found at https://cdn.intechopen.com/pdfs-wm/28316.pdf)

Harari Y (2015) Homo deus—a brief history of tomorrow. Harvill Secker, London

Hawking S, Mlodinow L (2010) The grand design. Bantam Books, New York

Heisenberg W (1971) Physics and beyond. HarperCollins, New York (German original: Der Teil und das Ganze: Gespräche im Umkreis der Atomphysik. Piper, München 1969)

Heisenberg W (1974) Across the frontiers. Harper & Row, New York (German original: Schritte über Grenzen. Piper, München 1971)

Hoffmann B, Dukas H (1981) Albert Einstein: the human side. Princeton University Press, Princeton and New Jersey

Hümmler HG (2017) Relativer Quantenquark (Relative quantum nonsense). Springer, Heidelberg

Isaacson W (2008) Einstein: his life and universe. Simon & Schuster, New York

Jaeger L (2015) Die Naturwissenschaften – Eine Biographie (Science: a biography). Springer, Heidelberg

Jaeger L (2017a) Supermacht Wissenschaft: Unsere Zukunft zwischen Himmel und Hölle (Superpower science—our future between heaven and hell). Gütersloher Verlagshaus, Gütersloh

Jaeger L (2017b) Wissenschaft und Spiritualität – Zwei Wege zu den großen Geheimnissen Universum. Leben, Geist, Springer (Science and spirituality—two paths towards the great secrets of the universe, life, and mind), Heidelberg

Jammer M (1995) Einstein und die Religion (Einstein and religion). Universitätsverlag Konstanz, Konstanz

Kahneman D (2012) Thinking, fast and slow. Farrar, Straus and Giroux, New York

Kaku M (2011) Physics of the future: how science will shape human destiny and our daily lives by the year 2100. Doubleday, New York

Kaku M (2014) The future of the mind. Doubleday, New York

Kant I (1781) Critique of pure reason (German original: Kritik der reinen Vernunft, Hartknoch, Riga, Ausgabe der Königlich Preußischen Akademie der Wissenschaften, Berlin 1902)

Kant I (1790) Critique of judgement (German original: Kritik der Urteilskraft, Hartknoch, Riga, Ausgabe der Königlich Preußischen Akademie der Wissenschaften, Berlin 1902)

Kelly J (2013) Smart machines: IBM's Watson and the era of cognitive computing. Columbia University Press, New York

Kepler J (1997) Mysterium Cosmographicum and Harmonices Mundi. English translation by Aiton EJ, Duncan AM, Field JV, The harmony of the world. American Philosophical Society, Philadelphia (original edition: Harmonices Mundi 1619)

Kohl C (2012) Nagarjuna and quantum physics: Eastern and Western modes of thought. AV Akademikerverlag, Saarbrücken

Kunmar M (2008) Quantum: Einstein, Bohr, and the great debate about the nature of reality. Norton & Company, New York

Kurzweil R (2005) The singularity is near: when humans transcend biology. Penguin Books, New York

Laughlin R (2007) A different universe: reinventing physics from the bottom down. Basic Books, New York

Lukrez (1473) De rerum natura (English translation: On the nature of things, translated by Rouse WH (1924), Revised by Smith MF (1992)). Harvard University Press, Cambridge)

Lüst D (2011) Quantum fishes: string theory and the search for a theory of everything (German edition: Quantenfische – Die Stringtheorie und die Suche nach der Weltformel). C. H. Beck, München

Mann F, Ch. (2017) Es werde Licht (Let there be light). S. Fischer, Frankfurt

McEvilley T (2002) The shape of ancient thought—comparative studies in Greek and Indian philosophy. Allworth Press, New York

McGrayne SB (2012) The theory that would not die: how bayes' rule cracked the enigma code, hunted down Russian submarines, and emerged triumphant from two centuries of controversy. Yale University Press, New Haven

McMahon D (2007) Quantum computing explained. Wiley, Hoboken

Milburn GJ (1997) Schrodinger's machines: the quantum technology reshaping everyday life. W. H. Freeman, New York

Müller V (ed) (2016) Fundamental issues of artificial intelligence. Springer, Berlin

Nagarjuna (trans. Garfield J) (1995) The fundamental wisdom of the middle way: Nāgārjuna's Mūlamadhyamakakārikā. Oxford University Press, Oxford

Newton I (1687) Philosophiae Naturalis Principia Mathematica (English translation: Mathematical principles of natural philosophy. Encyclopædia Britannica, London)

Nyanatiloka (1946) The word of Buddha. Buddhist Publication Society, Kandy Celon

Penrose R (1989) The emperor's new mind: concerning computers, minds, and the laws of physics. Oxford University Press, Oxford

Penrose R (1995) Shadows of the mind: a search for the missing science of consciousness. Oxford University Press, Oxford

Planck M (1949) Vorträge und Erinnerungen (Lectures and reminiscences). Hirzel, Stuttgart

Platon (2004) Theaitetos (English translation in Focus Philosophical Library)

Platon (2006) Politea (English: The republic, translated by Allen RE. Yale University Press, New Haven)

Platon (2009) Phaidon (English: Phaido, translated by Gallop D. Oxford University Press, Oxford)

Poincaré H (1914) Science and method. Thomas Nelson and Sons, London (French original: Science et méthode. Flammarion, Paris 1908)

Rae A (2006) Quantum physics: a beginner's guide. Oneworld Publications, London

Randall L (2012) Knocking on heaven's door: how physics and scientific thinking illuminate our universe. Vintage, New York

Rid Th (2016) Rise of the machines: the lost history of cybernetics. Scribe UK, London

Ross A (2016) The industries of the future. Simon & Schuster, New York

Schmidt E, Cohen J (2013) The new digital age. Alfred E. Knopf, New York

Schrödinger E (1944) What is life? The physical aspect of the living cell. Cambridge University Press, Cambridge

Schwab K (2016) The fourth industrial revolution. Portfolio Penguin, London

Shiller R (2012) Finance and the good society. Princeton University Press, Princeton

Singh S (2005) Big bang: the origin of the universe. Harper Perennial, New York

Smith A (1776) An inquiry into the nature and causes of the wealth of nations (The wealth of nations). W. Strahan and T. Cadell, London

Smolin L (2009) The trouble with physics: the rise of string theory, the fall of a science and what comes next. Penguin, London

Smolin L (2013) Time reborn. Harcourt Publishing, New York

Stukeley W (1752/1936) Memoirs of Isaac Newton's life. Royal Society Library, London; Taylor & Francis, Hier nach

Weinberg S (1992) Dreams of a final theory: the scientist's search for the ultimate laws of nature. Pantheon, New York

Weinberg S (2001) Facing up: science and its cultural adversaries. Harvard University Press, Cambridge

Whitaker A (2012) The new quantum age—from Bell's theorem to quantum computation and teleportation. Oxford University Press, Oxford

Zeh H (2003) Decoherence and the appearance of a classical world in quantum theory. Springer, London

Zeilinger A (2010) Dance of the photons: from Einstein to quantum teleportation. Farrar, Straus and Giroux, New York

Zeilinger A et al (2000) The physics of quantum information: quantum cryptography, quantum teleportation, quantum computation. Springer, London

Index

Printed in the United States
By Bookmasters